The Economic and Social Regulation of Public Utilities

Utilities have long been essential for societies, supplying basic services for nations, organizations and households alike. The proper functioning and regulation of utilities is therefore critical for the economy, society and security. History provides an invaluable insight into important issues of the economic and social regulation of utilities and offers guidance for future debates. However, the history of utility regulation – which speaks of changing, diverse and complex experiences around the world – was sidelined or marginalised when economists and policy-makers enthusiastically embraced the question of how to reform the utilities from the 1970s. This book examines in depth the complex regulation and deregulation of energy, communications, transportation and water utilities across Western Europe, the United States, Australia, Brazil and India. In each case, attention is drawn to the changing roles of the state, the market and firms in the regulation, organization and delivery of utility services.

This book was originally published as a special issue of *Business History*.

Judith Clifton is Professor of Economics at the University of Cantabria, Spain.

Pierre Lanthier is Professor of Economic History at the Université du Québec à Trois-Rivières, Canada.

Harm Schröter is Professor of Economic History at the University of Bergen, Norway.

The Economic and Social Regulation of Public Utilities

An International History

Edited by
Judith Clifton, Pierre Lanthier and Harm Schröter

Routledge
Taylor & Francis Group
LONDON AND NEW YORK

First published 2013
by Routledge
4 Park Square, Milton Park, Abingdon, Oxon OX14 4RN
605 Third Avenue, New York, NY 10017

Routledge is an imprint of the Taylor & Francis Group, an informa business

This book is a reproduction of the journal *Business History*, volume 53, issue 5. The Publisher requests to those authors who may be citing this book to state, also, the bibliographical details of the special issue on which the book was based.

British Library Cataloguing in Publication Data
A catalogue record for this book is available from the British Library

ISBN 13: 978-0-415-62298-1 (hbk)

Typeset in Times New Roman
by Taylor & Francis Books

Publisher's Note
The publisher would like to make readers aware that the chapters in this book may be referred to as articles as they are identical to the articles published in the special issue. The publisher accepts responsibility for any inconsistencies that may have arisen in the course of preparing this volume for print.

Contents

Citation Information

The following chapters were previously published in the journal *Business History*, volume 53, issue 5 (August 2011). When citing this material, please use the article's original page numbering, as follows:

Chapter 2
Geo-politics versus market structure interventions in Europe's infrastructure industries c. 1830–1939
Robert Millward
Business History, volume 53, issue 5 (August 2011) pp. 673-687

Chapter 3
Infrastructure and nation building: The regulation and financing of network transportation infrastructures in Spain (1720–2010)
Germà Bel
Business History, volume 53, issue 5 (August 2011) pp. 688-705

Chapter 4
States and start-ups: Public competitors in Australian communications
Jock Given
Business History, volume 53, issue 5 (August 2011) pp. 706-722

Chapter 5
How politics, economics, and institutions shaped electric utility regulation in the United States: 1879–2009
William J. Hausman and John L. Neufeld
Business History, volume 53, issue 5 (August 2011) pp. 723-746

Chapter 6
The 3 Rs: Regulation, risk and responsibility in British utilities since 1945
Martin Chick
Business History, volume 53, issue 5 (August 2011) pp. 747-760

Chapter 8
From national monopoly to multinational corporation: How regulation shaped the road towards telecommunications internationalisation
Judith Clifton, Francisco Comín and Daniel Díaz-Fuentes

Business History, volume 53, issue 5 (August 2011) pp. 761-781

Chapter 9

Public utilities and private initiative: The French concession model in historical perspective
Dominique Barjot
Business History, volume 53, issue 5 (August 2011) pp. 782-800

Chapter 1 in this book is an extended version of the introductory article originally published in *Business History*, volume 53, issue 5 (August 2011), pp. 659–672. To view this article in its original format, please refer to: http://www.tandfonline.com/doi/abs/10.1080/00076791.2011.599592

Notes on Contributors

Dominique Barjot is Professor of Modern Economic History at the Paris-Sorbonne (Paris IV) University, France. He was Director of the Humanities of the French Ministry of Research (2003–07), Chairman of the French Economic Historians Association (1998–2000), Deputy Director of Humanities and Social Sciences of the National Scientific Research Center 1994–97), and member of the Board of Standing Committee of Humanities of the European Science Foundation (1994–98).

Germà Bel is Professor of Economics at Universitat de Barcelona (GiM-IREA), Spain, and Guest Professor at Barcelona Graduate School of Economics.

Martin Chick is Reader in Economic and Social History at the University of Edinburgh, Scotland.

Judith Clifton is Professor of Economics at the University of Cantabria, Spain.

Francisco Comín is Professor of Economic History at the Universidad de Alcalá de Henares, Spain.

Daniel Díaz-Fuentes is Professor of Economics at the Universidad de Cantabria, Spain.

Jock Given is the author of *Turning off the television: Broadcasting's uncertain future* and articles about communications in Telecommunications Policy, *Historical Journal of Film Radio and Television*, *Info*, *Historical Records of Australian Science*, *Media International Australia*, *Telecommunications Journal of Australia* and *Inside Story*: http://www.inside.org.au.

William J. Hausman is Professor of Economics at the College of William & Mary in Williamsburg, Virginia, USA. He is the co-author (with Peter Hertner and Mira Wilkins) of *Global electrification: Multinational enterprise and international finance in the history of light and power, 1878–2007* (2011).

Pierre Lanthier is Professor of Economic History at the Université du Québec à Trois-Rivières, Québec, Canada.

Robert Millward's recent publications include *Private and public enterprise in Europe: Energy, telecommunications and transport c.1830–1990*, a chapter on population and living standards 1914–45 in the new (2010) *Cambridge economic history of modern Europe 1700–2000*, an article in *Business History* (2010) on privatisation and a chapter on business and the state in the *Oxford handbook of business history.*

John L. Neufeld is Professor of Economics at the University of North Carolina, Greensboro, USA. He is the author of numerous articles on the history of the electric utility industry.

Harm Schröter is Professor of Economic History at the University of Bergen, Norway.

The economic and social regulation of public utilities: An international history

Judith Clifton, Pierre Lanthier and Harm Schröter

In recent years, path-breaking studies which map out the evolution of the regulatory frameworks which have governed public utilities over the nineteenth and twentieth centuries have been produced by business and economic historians. Of particular importance is the volume by William J. Hausman, Peter Hertner and Mira Wilkins (2008), which focuses on the financing of the electrification process around the world from the 1870s to the present, and that by Robert Millward (2005), which analyses regulation of energy, telecommunications and transport in Western Europe from the 1830s to the 1990s. These studies coincide in their argument that utility regulation has taken many diverse forms in different geopolitical, economic and sectoral contexts, but that it is of intrinsic interest to other business and economic historians to attempt to gauge overall patterns of utility regulation over time, wherever possible. Both studies argue that utility regulation can be organised, generally speaking, into three successive "waves", whilst also recognising that many alternative paths to utility organisation and regulation were pursued simultaneously. In general terms, they argue that a "first wave" can be identified, particularly when infrastructure is initially constructed, where the high amount of investment required in sectors such as electricity, telecommunications and railways, as well as perceptions of great risk, made private alliances of entrepreneurs and families with banks and holdings essential as financiers. The state was involved in its capacity of adjudicating and granting rights of way, as well as regulating prices and service quality (Millward, 2005). Nevertheless, on occasions, the state acted as financier, especially when there were shortages, as in the case of the railways in France and Scandinavia, but also for strategic reasons, such as in Belgium (Millward, 2005: 59).

The following "wave" of infrastructure regulation occurred gradually, from the end of the nineteenth century, being consolidated from the interwar period, and was characterized by the growing role of the state in infrastructure finance, management and, often, ownership. During this period, regional and national networks were becoming technologically possible and, as new nation states were emerging, the importance of national security questions intensified (Millward, 2005: 91). Focusing on electricity, Hausman, Hertner and Wilkins (2008: 31) provide impressive quantitative data on ownership to show a process of what they call "domestication" rather than the more commonly used "nationalisation" for this period, since, though the state did not always assume ownership in all countries, the role of international finance in infrastructure was substantially reduced world-wide. At the same time, the state increased its involvement in utility regulation. Then, just as this stage was nearing completion, during the

1970s, the third wave began, which bore much in common with the first wave as regards the increased role of the domestic and foreign private sector and of market forces. One fascinating difference though, comparing the first and third stages is that, now, it was the public utility enterprises themselves who were the protagonists of this development. The upshot of this was that a significant number of former utility monopolies – particularly those based in the European Union – emerged in a short time span to become some of the world's largest Multinational Corporations (Clifton, Comín and Díaz-Fuentes, 2007). These three waves of infrastructure regulation clearly reflect broader changes in the world economy and economic policy, whereby ideas, ideologies and policy-makers assign different weights to the participation of the state and the market in governance (Toninelli, 2000). In the contemporary period, we are still in the third, market-driven phase, though there is by now substantial evidence that all is not perfect. During this period, there have been important cases of successes in infrastructure reform, but also instances of privatisation and regulatory failures and reversals, as well as issues of under-investment in infrastructure around the world due to cream-skimming, particularly affecting developing regions (UNCTAD, 2008). Moreover, in the context of the financial and economic crisis, policy-makers in the United States and Europe have called for more investment and better regulation of infrastructure as part of a possible exit strategy.

The central argument we wish to make here, in the light of this mixed evidence during this third wave, is that history can provide invaluable insights into important issues of utility regulation, and provide lessons towards future debates, but that history was sidelined or marginalized when economists and policy-makers enthusiastically embraced the question of how to reform the utilities from the 1970s. Each of the chapters included in this volume reflects this same argument, in different ways. Utility regulation from the 1970s tended to be founded on assumptions or ideologies which directly countered those underpinning regulation in the previous phase. In particular, proponents of reform in the third phase tended to argue that the foundations structuring previous regulation were increasingly irrelevant and obsolete. More recently, there has been recognition that the dismissal of previous concerns about the special role of utilities in society and the economy went too far, and there has been some backtracking or reconsideration of the issues, as shall be discussed. Before turning to the discussion of the chapters, the main tenets underpinning, and some of the practical experience of, the second and third waves of regulation are discussed.

From the end of the nineteenth century, the state's role gradually increased in the management and ownership of utilities. This intervention was justified by a complex mix of economic, military, political and social arguments. Firstly, from the economic perspective, utilities were understood as exhibiting particular features such as market failures (particularly due to the problem of the natural monopoly), high sunk costs and network economies. It was therefore sustained that services of this nature should be provided by a single enterprise or organisation, such as an electricity central board, at the local or national level. Secondly, from the military perspective, utilities were responsible for providing communication, energy and transportation, all critical services in the defence of the nation. In the context of the aftermath of two, recent, world wars, the vital physical as well as psychological role played by infrastructure in defending the nation were still at the forefront of policy-makers' minds (Taylor, 2003). Ultimate government control over infrastructure services, whether through ownership and/or regulation, was therefore seen as essential. Thirdly, from the political perspective, a government's interest in regulating the utilities stemmed from its interests in influencing network development as geopolitical spaces themselves changed. Prior to the construction of modern Nation States, it may have been adequate to provide services provided by utilities on a "club" basis. The lack of connections between networks served to defend providers against competition, but also helped to defend designated areas from attack, by restricting movement across borders or spaces, for example, by blocking easy access to local networks. However, as Nation States came into their own, particularly from the nineteenth century, it became more important to integrate citizens into this national space. State regulation was often aimed at encouraging, or forcing, multiple, individual enterprises to inter-connect across territory to forge a national network. This emerging national network took various forms: in countries such as France or Spain, emphasis was on imposing a single political centre, where the network hub would be established, whilst the peripheries were like spokes on a wheel, connected to the hub, but less so to each other (Flichy, 1995). In contrast, in federal systems, such as the United States or Germany, several hubs were established across the territory, and linked to their respective peripheries. In addition, state policy on utilities was often connected to their interest in linking the nation to their empire. States thus became managers and, often, owners of large technical systems (Bijker, Hughes and Pinch, 1989). A government was also well positioned to drive network development as rights of way often needed establishing. Finally, socially motivated influences on regulation of public utilities, understood as public services, included governments' will to shape national redistributive policies. General welfare considerations were prioritized over and above individual benefit, since the latter was associated with fragmenting society and regions, and of being regressive, potentially undermining national and social collective action (Reynolds, 2004). In the case of the utilities, policies included the subsidization of important loss-making utilities with more profitable ones, hence sustaining jobs and services for communities across the whole range of services, as well as cross-subsidization policies, where network take-up and use in lower socio-economic households or in geographical peripheries was facilitated through subsidized prices, extracted from more profitable parts of the service, such as services more usually used by richer segments of society or from the hub (Clifton, Comín and Díaz-Fuentes, 2003). The practical outcome of state intervention in utilities was varied. Despite this, towards the end of the second wave, utilities increasingly came under fire for inefficiency, corruption, poor service quality and sometimes, their generation of enormous losses (Toninelli, 2000). Momentum increased for a new approach to governing utilities, and views on how to do this converged during the 1970s. To sum up, in most countries around the world, it was held that the state should assume a predominant role in the regulation of utilities, very often as regards ownership, and nearly always, as regards its responsibilities as over-seer. But this approach was soon to change substantially.

With a view to addressing what had been identified as some of the key problems of utility performance in the second wave, proponents of reform stressed that new policies were needed to ensure utilities attained superior performance results, delivered better service quality and choice to users at lower prices (Kessides, 2004). Now, utility regulation would be supported by policies including privatisation, liberalisation and deregulation. The push to the new policy paradigm was partly fuelled by technological change, which was particularly significant in telecommunications due to convergence and digitalization, relevant in the cases of electricity, gas and transportation, though less so in sectors such as water. Technological change in the telecommunications sector, it was argued, eroded the argument that infrastructure was characterized by natural monopoly and economies of scale, since sunk costs were less significant and the market became more contestable (Bauer, 2010). Indeed, technological change in the telecommunications sector, particularly the increased importance of new customer premise equipment (fax, telex terminals, multiple telephone handsets and so forth) which had no claim to monopoly status, resulted in growing pressure from business groups in the mounting challenge to the telecommunications sector's continued monopoly privileges (Millward, 2005: 252).

But the change in policy paradigm was also associated with a renewed attention by policy-makers and scholars to particular economic and managerial theories, particularly those influenced by public choice, which held that publicly-owned bureaucracies represented an inevitable obstacle to efficiency due to the incentive problem (Osborne, 1993). To rectify this, services provided by utilities, just as other industrial goods and services, should be increasingly subject to pressures from competition from the market, even if unbundling policies to separate competitive and non-competitive elements were required, or the government had to find other means of promoting competition in markets that were essentially monopolistic. Whilst government ownership of utilities had been the norm around most of the world, policy recommendations from the international economic institutions now widely recommended the virtues of private ownership. Exposing the former monopolies to private-firm styles of management in a competitive or non-competitive environment was proposed as the solution to improving their efficiency and performance by many international economic organisations, including the World Bank and the OECD (Clifton and Díaz-Fuentes, 2011). The beneficiaries, it was claimed, would ultimately be the consumers, who would obtain a greater choice of services, at a better quality and a lower price. Welfare was understood as an aggregation of individual benefits, rather than as a social outcome. Gradually, the so-called "Washington Consensus" (Williamson, 2004) emerged about this new direction of policy. Arnt Aune (2000) traced the ways in which policies based roughly on neoclassical economic theory were diffused from think tanks and specific universities based in the United States to government departments and the mass media around the world. Stiglitz (2003) later criticized this one-fits-all approach to policy as amounting to an act of faith when he described it as "market fundamentalism". Others have observed how the push toward deregulation was presented as being "one-way", meaning that the benefits of deregulation were supposed to be so sure that reversals, in the form of re-nationalisations, contracting back in and re-regulation, were not predicted (Hefetz and Warner, 2004). The fact that reversals had been common across the history of utility regulation seemed to have been forgotten (Comín and Díaz-Fuentes, 2005). What is clear is that the promoters of the reform of utilities adopted an *aspecific* and an *ahistorical* approach (Clifton, Lanthier and Schröter, 2011). Utilities were thought to be just as suitable a candidate for reform as other business from industrial sectors. Previous justifications for utility regulation based on their specificities and complexities had been largely dismissed by those enacting reform from the 1970s onwards, so that the essential roles of utilities in economic, technological, political and social terms were underestimated, or overlooked, whilst an oversimplified vision of the future of policy on utilities was promoted.

Hindsight, provided by over three decades of experience of utility privatisation, liberalisation and deregulation, accompanied by a growing body of empirical studies on the effects of reform, reveals that the experience of utility reform across different sectors and countries world-wide was mixed. Utilities, it turned out, were a much more complex set of objects to reform than had been assumed by proponents of reform in the third wave (OECD, 2002). If utility deregulation was complex in the developed countries, it often proved even more problematic in the developing nations where discontent with utility reform inspired by the Washington Consensus increased dramatically (Checci, Florio and Carrera, 2009). Antonio Estache, a leading World Bank economist specializing in utility reform, stated that "the most dramatic lesson the international infrastructure community may have learned is humility", recognizing the limits and weaknesses of policy advice from the international organisations to developing countries (Estache, 2006). Admitting the difficulties utility reform was having, the World Bank even began commissioning reports into emerging, or re-emerging problems, such as massive corruption in the privatised utilities (Kenny and Soreide, 2008).

Problems caused by infrastructure reform were multiple and complex. Privatisation policies brought in sorely needed investors, but inevitably on an uneven scale around the world, since the location, sector and related prospects for profit, mattered. In other words, profit-oriented private firms, sometimes in collaboration with governments, cream-skimmed infrastructure projects. So, whilst the sale of former telephone monopolies in both developed and developing regions, such as BT in the United Kingdom and TELMEX in Mexico, blazed the trail of ambitious privatisation programmes (Clifton, 1999), it proved much more complex to attract inward Foreign Direct Investment into certain utility sectors in poorer regions. Indeed, under-investment into infrastructure was deemed of such importance that the United Nations Conference on Trade and Development (UNCTAD) dedicated its annual *World Investment Report* to this issue (UNCTAD, 2008). Neither was privatisation one-way: in the United Kingdom, where privatisation had been early, deep, and rapid, reversals occurred, as in the case of the re-nationalisation of British railways. Privatisation reversals also occurred across the developing world, for instance, in the water sector (Hall and Lobina, 2008) as well as in the United States, especially when cost-savings failed to materialize, whilst alternatives to privatisation grew, such as inter-municipal cooperation and government entrepreneurialism (Warner and Hebdon, 2001).

As regards introducing competition into the utilities, this proved, again, easier in some sectors, such as telecommunications, than in others, such as energy and water. In the context of the European Single Market, concern about the concentration of market power has increased in recent years (Clifton, Díaz-Fuentes and Revuelta, 2010). Stephen Thomas (2003) argued that market integration was leading, instead of to greater competition, to greater concentration, and labelled the remaining energy companies in the European Union the "seven brothers", though even this irony may have been over-optimistic since only five or six major companies now remain. The policy of unbundling proposed to separate the organisation of firms into two constituent parts; those that could be exposed to competition in some form, and those that could not. Hence, in the case of electricity, generation and supply to final customers could become competitive, but high-voltage transmission and local distribution could not (Gonenc, Maher and Nicoletti, 2001). Another problem was associated with regulation. What was supposed to be "deregulation" actually turned out to be "re-regulation", in that numerous independent regulatory agencies were established. It was thought initially that these bodies could be transitory, until the industry had been consolidated. Instead, this gave rise to a complex, seemingly permanent, new set of regulatory arrangements (Thatcher and Coen, 2008).

Downplaying – or ignoring – the lessons of history on the complexities of utility regulation came at a cost, most dramatically, during the 2000s, when new forms of terrorism used public utilities to organise (mobile telephony, internet) and attack (postal services, airlines, metros and buses) citizens and organisations around the world. The way in which governments believed their control over utilities had become less important from the 1970s has since been subject to re-examination. In the United States, following September 11, the Committee on Foreign Investment in the United States introduced new restrictions on inward foreign investment into the so-called "strategic sectors", including communications and transportation infrastructure. In the European Union, similar concerns about identifying and protecting "critical infrastructure" are re-surfacing in debates among policy-makers (EC, 2006), whilst France and Germany have introduced new measures at the national level to restrict investment (Clifton and Díaz-Fuentes, 2010). While policy-makers were far from advocating a return to the second wave of utility regulation, there was a generalised perception that proponents of the third wave had gone too far in their claims that utilities could be regulated just as any other sector of the economy. This is the starting point from which the chapters in this volume can be read. Each chapter examines and rethinks utility regulation from a long-term, international perspective, with special emphasis on the Western world. Authors were requested to place emphasis on the historical complexities of utility regulation, in terms of instruments, objectives and results, across sectors and countries, with a view to extract insights and lessons from history for the appropriate regulation of utilities in the future.

In this light, a brief synthesis of the chapters included here is provided. Robert Millward, Germà Bel and Jock Given, authors of the first three chapters in this volume, adopt a long-term perspective on how utilities have been regulated with a view to determining what relative weight political and economic factors played in the different institutional settings. Millward's chapter is comparative, across multiple European countries and sectors; Germà Bel focuses on transport infrastructure in Spain, including road and rail, whilst Given analyses the communications sector in the context of Australia. Interestingly, all three authors coincide that, though the economic characteristics of networks mattered, as did technological change, political influence over regulation was decisive. Millward's chapter can be read as part of his ongoing work to synthetically explain the commonalities and differences in the regulation of the utilities across Western Europe from the middle of the nineteenth century to the present, with the aim of identifying which factors were most influential in shaping regulation (Millward, 2005). Focusing on the period 1830–1939, Millward stresses how a mix of economic, military, political and social factors shaped utility regulation in Western Europe, but asks in particular why state intervention into the utilities was more intense across Continental Europe than in Britain during this time. His answer is that Continental Europe consisted of a set of contiguous and often hostile nation states, and that the perception of the geo-political importance of utilities predominated over perceptions about their economic characteristics, such as the problem of natural monopoly. In contrast, in Britain, an island economy with a strong navy and merchant fleet, disruptions to the communications and trade systems were less of a concern than in the Continent. Moreover, as a first-mover in shipping, coal, and international telegraphy, the informal management of utilities was deemed to be sufficient, and preferable to outright control. Thus, he argues that this heightened concern about the role of utilities in defence in the Continent explains governments' more intensive intervention vis-à-vis their British peer. Meanwhile, the main thrust of state intervention was qualitatively different: in Britain, the emphasis was on price and service quality; in contrast, in the Continent, regulation was more related to security issues.

Continuing with the question of the relative weight of the importance of political versus economic rationales as the predominant logic driving the design of infrastructure regulation, Bel argues that the evolution of infrastructure policy in Spain from the eighteenth century to the present was dominated by political interests, over and above commercial or other economic interests. The fundamental explanation, he argues, is the drive to centralize the Spanish mainland with Madrid as its hub. On these grounds, subsidies were used to sustain this centralising policy, despite projects lacking at times solid commercial or economic justifications. A long-term drive to nation-building is examined for road building during the eighteenth century, and railway construction and expansion during the nineteenth century. During the twentieth century, motorway expansion was subsidised to enhance Spain's radial organisation, again, with Madrid at the centre, repeating the pattern applied to the rail system. Finally, this century, the high-speed train, whose first branch linked Madrid to Seville (home town of former President Filipe González!) cannot be justified by cost-benefit analysis, but rather, by political interests. Bel critiques this policy observing that Spain has the second best high-speed train network in the world after China, though it has a much smaller percentage of users than its French neighbours.

Some of the deepest reform of utilities is found in the communications sector. Jock Given's chapter also seeks to identify the relative weight of economic and political factors driving utility regulation, but does so from the perspective of the individual firm. His chapter comparatively explores the establishment, evolution and eventual demise of three communications enterprises in Australia: Pacific Cable, which opened in 1912; Amalgamated Wireless Australasia, which launched wireless telegraphy services in 1927; and AUSSAT, a satellite system established in 1985. In each enterprise, the state participated in significant, but different, ways. Given's general argument in all cases is that the state got involved in regulation above all for a mix of political and pragmatic reasons. The first two projects were international in reach, and were predominately driven by empire-building ambitions; in particular, improving communications was seen as a critical means of closing the huge distance between Britain, its territories, and Australia. The third project was regional in nature, and was motivated by nation-building objectives. In all three cases, the enterprises were established in order to compete with an incumbent. State participation was crucial in that this was seen as the only way to forge competition. Though, in all three cases, the government insisted that the generation of competition would be a profitable business, reality indicates they were motivated by other policies, such as trying to bring down prices, setting up alternative infrastructures, and making more widely available services. All three enterprises came to their eventual demise, which usually involved Cable & Wireless; tellingly, an enterprise which merged technologies retained clear advantages over the single-technology firms.

Utilities providing electricity have been subject to deep, though complex and sometimes, controversial reform, whilst less has been achieved by policy-makers in the field of water. The next three chapters, by William Hausman and John Neufeld, Martin Chick and Pierre Lanthier, respectively, vividly highlight the complex long-term evolution of the political economy of utility regulation in different settings, the United States, Britain and France, and Brazil and India. These chapters contextualise analysis by using the rich historical background of each economy in order to contrast how the deregulation and restructuring of utilities from the 1970s ended up far from being a "panacea". Hausman and Neufeld focus on analysing the organisation of the electricity sector from its beginnings to the present in the United States, deploying explanations for regulation based on the economic and technological characteristics of the electricity sector, including the processes of generating and distributing electricity, as well as political features of the country. Against this background, the authors discuss one of the most controversial of deregulatory experiments; that pioneered by the Californian government, which, in retrospect, has been classified by most observers as a "big mistake". Towards the end of their chaper, the authors discuss what went so wrong in the California debacle. They find blame in abuses of market power, corruption and flawed (de)regulation, among other factors. Then, they show the consequences of these occurrences in the way that, from 2001, electricity utility restructuring has been frozen or reconsidered in a total of twenty-nine states; eight have still not embarked on reform, and a total of only eleven states have opted to continue. That deregulation was no panacea is a common thread with Chick's chapter, which analyses the ways in which three concepts, regulation, risk and responsibility, shifted across the twentieth century in the organisation of two utilities, electricity and water, in the United Kingdom. Drawing on his previous work, Chick contrasts the ways in which price and rates of return regulation was computed for nationalised industry and then during the run-up to and aftermath of privatisation (Chick, 2007). His discussion is informed by an interest in the socio-economic consequences of such a shift, with the result that by the 1990s, restructuring meant that, whilst prices for electricity to consumers increased, those for industrial users fell, a trend confirmed by Florio (2004). Moreover, increased electricity prices fell harder on the elderly and the poorer whilst increased water prices occurred on a regional basis. Privatisation and liberalisation caused, or exacerbated, fuel and water poverty, Chick argues. The new regulatory paradigm re-introduced risk and uncertainty into industries that had not really endured these for decades, and inadequate attention was paid to how this new risk would affect prospective sunk investment. Finally, he argues that a new approach was taken as regards responsibility: poverty came to be increasingly dealt with as an outcome (such as the "cold weather payment scheme" for pensioners), rather than through policies seeking redistribution. Pierre Lanthier provides a key chapter to this edited volume by turning his attention to two of the key emerging countries in the contemporary period. The cases of Brazil and India are both ones where huge future electricity demand has required – and continues to need – massive development of the industry, facilitated by inward FDI. In both cases, Lanthier shows how national preferences and the characteristics of the insertion of these countries in the world economy strongly influenced the ways in which the so-called "World Bank" model of regulation was finally adopted. He also charts the distinctive political, social and economic reasons to explain why initial reform plans either failed to be implemented, or were changed half-way through.

From the 1990s onwards, some utilities, many of which were state-owned and run as monopolies for decades, emerged rapidly to occupy leading positions in the ranking of world Multinational Corporations. The two final chapters in this volume take up this rather surprising and still under-researched development. European-based utilities are leading the pack around the world, largely, as a consequence of service liberalisation in the Single Market. The chapter by Clifton, Comín and Díaz-Fuentes seeks to identify whether and how the regulatory framework that governed utilities over the long term could be an important part of the story explaining their internationalisation. To do so, they examine the long-term regulatory experience and more internationalisation experiences of two large telecommunications firms: BT and Telefónica. Both are based in large European countries but, whereas BT became one of the least internationalised of all large European telecommunications incumbents, Telefónica became the most internationalised of all. The authors find explanations for the divergent internationalisation strategies in historical differences in ownership, management style, capital access and exposure to liberalisation. Telefónica was unique in the European context, born a private company and controlled by a foreign Multinational, ITT. Its nationalisation was never completed, and its management was closer to that of a private firm than the traditional General Office of Post and Telegraph, and it enjoyed fluid capital access. Finally, despite privatisation and liberalisation elsewhere, Telefónica's early acquisitions in Latin America were conducted from its privileged monopoly position, whilst, post liberalisation, the firm continued to enjoy a high market share. BT, in contrast, was organised under state ownership early on, and controlled by the public administration from the end of the century. BT was sold as an integrated monopoly in 1984, and the government had to sponsor a rival, Mercury, in their duopoly policy. Liberalisation in the UK was well ahead of the European average, and BT was not "pampered" by the government, as incumbents were in some Continental countries. Its strategy to pursue global alliances, rather than Foreign Direct Investment, failed time and time again; then, to protect eroding home markets, BT partially de-internationalised, shrinking, becoming the least international of the major European telecoms operators.

The final chapter, by Dominique Barjot, also explores the relationship between utility regulation and internationalisation over the long term in the case of France. Here, the author asks why this country has produced some of the contemporary world's most important and successful enterprises from the public works and public utilities sectors, including urban services, such as water, energy, transportation and other infrastructure. The key reason, he argues, lies in the distinct legal traditions and definitions that existed for centuries which defined public works and public utilities. This legal tradition allowed for, and actually encouraged, substantial private sector involvement. In this context, the private sector participated in a variety of ways, along with the public sector, in different forms, towards the construction, operation and management of these services. In other words, the French legal system stimulated the development of "network capitalism" by establishing financial and managerial systems which allowed for the long-term participation of private capital in the development of major infrastructure projects. After presenting the historical evolution of the legal framework governing infrastructure and the role of private capital therein, he explores the development of four, major Multinational Corporations based in France: Vinci, Bouygues, Générale des Eaux and Lyonnaise des Eaux. Their dramatic internationalisation particularly from the 1990s was facilitated also by the growing role of cities in contemporary capitalism. As the CEO of Veolia Environnement claimed recently in an interview, he "dreams of US urban cities" – the growth of huge, complex urban cities is precisely where utilities can sell their decades of specialist management experience (Amiel, 2011). Indeed, in the face of greater urbanisation, the shift towards economic growth to the East and the South, depletion of natural resources and issues of global climate change, new forms of terrorism and war, the sustainable, safe and fair regulation of utilities constitutes a key, future task for governments and business around the world. Let us hope the historical idiosyncrasies of infrastructure are not marginalised in the next wave of their regulation.

Utilities over the long-term: A scholarly network

This last section contains a final few words about the scholarly network around utilities over the long-term. From the 1990s, we were working in two, parallel, groups. Judith Clifton was introduced to the World Economic History Conference (WEHC) network by Francisco Comín, who had served previously as General Secretary of the Spanish Economic History Association, and Daniel Díaz-Fuentes, during the 1990s. This group was working together on the privatisation and nationalisation of public enterprises in the European Union. Towards the end of that decade, they applied to a pre-session of the WEHC to be held in Trois-Rivières, Canada, organised by Pierre Lanthier. In the event, it turned out that the pre-session was held on 11–12 October 2001, just one month after September 11th. In truth, the idea of taking a transatlantic flight was not a very attractive one, especially as Clifton had to travel with her new-born son. Temptation to call everything off was resisted and then finally celebrated due to the excellent organisation of the Trois-Rivières pre-session. Not only did Pierre put participants up in the most exquisite of hotels, he also fed us with the best of *haute cuisine*. Conference attendees included Dominique Barjot, Claude Bellavance, Alain Beltran, Martin Chick, Lina Gálvez, Pierre Mounier-Kuhn, H. Vivian Nelles, Pedro Pablo Ortúñez, Aron Shai, Pier Angelo Toninelli, Nuno Valerio and Warren Young. The result of this most enjoyable pre-session was presented at the WEHC in Buenos Aires in 2002 (Chick and Lanthier, 2004). The expert for this session was Patrick Fridenson, and the time, energy and dedication he paid to our session marked the beginning of a strong academic and intellectual friendship. From this session two special issues were published in scientific journals: firstly, an issue edited by Alain Beltran, Martin Chick and Pierre Lanthier entitled 'Nationalisations et dénationalisations de l'électricité' in *Annales historiques de l'électricité*, no 1, Presses universitaires de France et Victoires éditions, June 2003 and, secondly, Martin Chick and Pierre Lanthier (eds), 'Nationalisations et dénationalisations' in *Entreprises et histoires*, 37 (3), ESKA, December 2004.

In parallel, another team was working on the long-term evolution of the process of electrification around the world, particularly, as regarded the changing ways of how this had been financed. William Hausman, Peter Hertner and Mira Wilkins organised a pre-session on that topic in Wittenberg, Germany, and then a session at the WEHC in 2002, followed by several more meetings, including one held in May 2003 on the invitation of EDF, organised by Dominique Barjot and Peter Hertner, and then in June 2003 a panel at the Business History Conference under the organisation of William Hausman. The outcome of these meetings was the important volume *Global Electrification: Multinational Enterprise and International Finance in the History of Light and Power, 1878–2007* (Hausman, Hertner and Wilkins, 2008) to which Dominique Barjot, Jonathan Coopersmith, Kenneth E. Jackson, Pierre Lanthier, H. V. Nelles, John L. Neufeld, Harm Schröter and Luciano Segreto, also contributed.

Because the utilities examined from the perspective of nationalisation and denationalisation were emerging as some of the world's leading Multinational Corporations from the 2000s, and that this had clear historical precedents, Judith Clifton, Francisco

Comín and Daniel Díaz-Fuentes applied for a session at the WEHC on the internationalisation of utilities. After session acceptance, they held a pre-session at the University of Cantabria inviting country experts to analyse this perhaps surprising development: Lena Andersson-Skog, Sean Barret, Carlos Bastien, Frans Buelens, Candra S. Chahyadi, Patrick Fridenson, Marina Klinova, Lina Gálvez, Carlos Marichal, William Megginson, Robert Millward, Ana Bela Nuñes, Tomas Pettersson, Jesús Salas, Pier Angelo Toninelli, Julien van den Broeck, Michelangelo Vasta, Nuno Valério and Hans Willem. After the successful pre-session, the main concern was that Germany was missing from the analysis, so Harm Schröter accepted to step in. At the same time, Schröter was organising his pre-session for WEHC in Milan on the emergence of a European enterprise. This was published as *The European Enterprise: Historical Investigation into a Future Species* (Schröter, 2008). Because work from the other project, on the internationalisation of utilities based in the European Union, could be interpreted as leading to a European-based firm, the interest of the two projects converged.

Positive reception of the session on the internationalisation of utilities led to the publication of *Transforming Public Enterprises in Europe and North America* (Clifton, Comín and Díaz-Fuentes, eds., 2007) after the WEHC in Helsinki, 2006. In addition, Mira Wilkins kindly recommended Judith Clifton and Daniel Díaz-Fuentes as experts to the preparation of the *World Investment Report*, which was to focus on infrastructure (UNCTAD, 2008). Here, they met Peter Buckley, John Dunning, Hafiz Mirza, Anne Miroux and Rajneesh Narula, among others, and helped towards the establishment of the UNCTAD-academic network where close contact with policy-makers in developing countries helped clarify new, urgent issues, such as under-investment and asymmetrical investment negotiation in infrastructure. Other, new colleagues and friends were made on the basis of common research interests, including Andrea Goldstein, OECD Investment Division, Karl Sauvant, of the Vale Columbia Center on Sustainable International Investment and Louis Brennan, Professor of International Business at Trinity College, Dublin, who is currently managing an important COST project on the rise of multinationals from the south and their impact on Europe.

For the WEHC in Utrecht in 2009, the issue of how utilities had been regulated, or de-regulated, over time became the question under study, particularly as it started to become apparent that the new regulatory model for utilities was proving far from perfect, and more difficult than many policy-makers had thought to apply. Indeed, partly as a result of regulatory failure, the financial and then economic crises made funding complex in order to prepare for the World Congress. The Plan B, the virtual pre-session, had to suffice on this occasion. Meeting at Utrecht in 2009, Patrick Fridenson and Robert Millward, the session's experts, coincided that the question of regulation and deregulation of utilities was still an open one, and that far more thinking needed to go into future regulatory models, particularly as regards the environment and the generational question. Thanks for the successful outcome are due, as usual, to many individuals, including the Clifton and Gardner families, who cared for Judith's children whilst we were working! After the session, over lunch, plans were made to split the papers into two, on the logic of those papers which dealt with one country and a single sector over a shorter time period, and those which dealt either with multiple sectors in one country or one sector across a large country over a considerable time-period. The first batch of papers would be organised into a special issue for *Entreprise et Histoire* (forthcoming); the latter, after a call for papers, to *Business History*, which was republished, in this volume.

Next, following the priorities established in the call for sessions by the WEHC organisers for Stellenbosch, South Africa, in summer 2012, our team will organise accepted session (number 109) that focuses on questions of international investment in infrastructure and the consequences for development based on experiences during the twentieth century. The ideas for this session were inspired by the experience with the team in the UNCTAD, particularly regarding issues of asymmetry between investors and host countries, problems of under-investment and so on. This ongoing work will no doubt be enriched by the participation by some of the members in a research project funded by the European Investment Bank on the "History of European Infrastructure Finance" 2011–2014, along with Hugh Goldsmith and colleagues from the universities of Milan, Bauhaus Universitat Weimar and Ludwig-Maximilians-Universitat München.

References

Amiel, G. (2011). Keeping our Cities Livable, Interview with Veolia Environnement's chairman and chief executive Antoine Frérot, *Wall Street Journal*, 31 January.

Arnt Aune, J. (2000). *Selling the Free Market: The Rhetoric of Economic Correctness*. Guilford Press.

Bauer, J. M. (2010). Changing roles of the state in telecommunications. *International Telecommunications Policy Review*, *17*, 127–145.

Beltran, A., Chick, M., & Lanthier, P. (eds) (2003). *Nationalisations et dénationalisations de l'électricité* in *Annales historiques de l'électricité*, no 1, Presses universitaires de France et Victoires éditions, June.

Bijker, W., Hughes, T., & Pinch, T. (1989), *The Social Construction of Technological Systems: New Directions in the Sociology and History of Technology*. MIT.

Checchi, D., Florio, M., & Carrera, J. (2009). Privatization discontent and utility reform in Latin America. *Journal of Development Studies*, *45*(3), 333–350.

Chick, M. (2007). *Electricity and Energy Policy in Britain, France and the United States*. Cheltenham: Edward Elgar Publishing.

Chick, M., & Lanthier, P. (2004) (eds.). *Nationalisations et dénationalisations. Entreprises et histoires*, *37*(3), ESKA, December 2004.

Clifton, J. (1999). Sindicatos y política en México: el caso de la privatización del Telmex. *Política y gobierno*, *2*, 1999, 407–440.

Clifton, J., Comín, F., & Díaz-Fuentes, D. (2003). *Privatization in the European Union: Public Enterprises and Integration*. Kluwer Academic Publishers.

Clifton, J., Comín, F., & Díaz-Fuentes, D. (2006). La privatización de las empresas públicas en la UE: ¿la vía británica o la senda europea? *Revista de Economía Mundial*, *15*, 121–153.

Clifton, J., Comín, F., & Díaz-Fuentes, D. (2007). *Transforming public enterprises in Europe and North America: Networks, Integration and Transnationalization*. Palgrave Macmillan.

Clifton, J., & Díaz-Fuentes, D. (2010). Is the European Union ready for FDI from emerging markets? In *Foreign Direct Investment from Emerging Markets: The Challenges Ahead*, Karl Sauvant with W. Maschek and Geraldine McAllister (eds.). London/New York: Palgrave Macmillan.

Clifton, J., Díaz-Fuentes, D., & Revuelta, J. (2010). The Political Economy of Telecoms and Electricity Internationalization in the Single Market. *Journal of European Public Policy*, *17*(7), 988–1006.

Clifton, J., & Díaz-Fuentes, D. (2011). The OECD and phases in the international political economy, 1961–2011. *Review of International Political Economy, 18*, 5, 552-569.

Clifton, J., Lanthier, P., & Schröter, H. (2011). Regulating and deregulating the public utilities 1830–2010. *Business History, 53*, 5, 659–672.

Comín, F., & Díaz-Fuentes, D. (2005). *La empresa pública en Europa*. Madrid: Síntesis.

Estache, A. (2006) *Infrastructure: A survey of recent and upcoming issues.* Paper presented to the World Bank, April 2006. http://siteresources.worldbank.org/INTDECABCTOK2006/Resources/Antonio_Estache_Infrastructure_for_Growth.pdf (last accessed 15 March 2011).

European Commission (2006). *Communication on a European Programme for Critical Infrastructure Provision, COM(2006) 786 final*, 12 December.

Flichy, P. (1995). *Dynamics of Modern Communications. The shaping and impact of new communications technologies.* London: Sage.

Florio, M. (2004). *The Great Divestiture: Evaluating the welfare impact of the British privatizations 1979–1997.* Cambridge, MA and London: MIT Press.

Galambos, L. (2000). State-owned enterprises in a hostile environment. In P. A. Toninelli (ed.), *The Rise and Fall of State-Owned Enterprise in the Western World*, Cambridge University Press.

Gonenc, R., Maher, M., & Nicoletti, G. (2001) *The Implementation and the Effects of Regulatory Reform: Past Experience and Current Issues*, OECD Economic Studies 32(11). OECD: Paris.

Hall, D., & Lobina, E. (2008). *Water Privatization*, Public Services International Research Unit Working Paper, http://www.psiru.org/reports/2008-04-W-over.doc last accessed 14 March 2011,

Hausman, W., Hertner, P., & Wilkins, M. (2008). *Global Electrification: Multinational Enterprise and International Finance in the History of Light and Power, 1878–2007.* New York: Cambridge University Press.

Hefetz, A., & Warner, M. (2004). Privatization and its reverse: Explaining the dynamics of the government contracting process. *Journal of Public Administration Research and Theory, 14*(2), 171–190. doi: 10.1093/jopart/muh012

Kenny, C., & T. Soreide (2008). *Grand Corruption in Utilities. Policy Research Paper 4805.* The World Bank, Washington.

Kessides, I. (2004). *Reforming Infrastructure. Privatization, Regulation and Competition.* World Bank Policy Research Report, World Bank/Oxford University Press.

Lipartito, K. (2000). Failure to Communicate: British Telecommunications and the American Model. In *Americanization and Its Limits: Responses to US Technology and Management in Postwar Europe and Japan*, edited by Jonathan Zeitlin and Gary Herrigel. Oxford University Press.

Millward, R. (2005). *Public and Private Enterprise in Europe: Energy, Telecommunications and Transport 1830–1990.* Cambridge: Cambridge University Press.

OECD (2002). *Restructuring Public Utilities for Competition.* OECD: Paris. http://www.oecd.org/dataoecd/6/60/19635977.pdf

Osborne, D. (1993). *Reinventing government: how the entrepreneurial spirit is transforming the public sector.* Plume Books.

Reynolds, L. (2004). Taxes, Fees, Assessments, Dues and the 'Get what you pay for' Model of Local Government. *Florida Law Review, 56*, 373–441.

Schröter, H. (ed.) (2008). *The European Enterprise: Historical Investigation into a Future Species.* Springer.

Stiglitz, J. (2003). *Globalization and its discontents.* New York: W.W. Norton and Company.

Taylor, P. (2003). *Munitions of the Mind. A history of propaganda from the ancient world to the present day.* Third Edition. Manchester University Press.

Thatcher, M., & Coen, D. (2008). Reshaping European regulatory space: an evolutionary analysis. Western European Politics, *31*(4), 806–836.

Thomas, S. (2003). The Seven Brothers. *Energy Policy, 31*(5), 393–403.

Toninelli, P. A. (ed.) (2000). *The Rise and Fall of State-Owned Enterprise in the Western World.* Cambridge/New York: Cambridge University Press.

UNCTAD (2008). *World Investment Report: The Challenge of Infrastructure.* UNCTAD: Geneva.

Warner, M. E., & Hebdon, R. (2001). Local Government Restructuring: Privatization and its Alternatives. *Journal of Policy Analysis and Management, 20*(2), 315–336.

Warner, M.E. (2008). Reversing Privatization, Rebalancing Government Reform: Markets, Deliberation and Planning. *Policy and Society, 27*(2), 163–174.

Williamson, J. (2004). The Strange History of the Washington Consensus. *Journal of Post Keynesian Economics, 27*(2), 195–206.

Geo-politics versus market structure interventions in Europe's infrastructure industries c. 1830–1939

Robert Millward

School of Arts, Histories and Cultures, University of Manchester, Manchester, UK

Throughout the nineteenth and twentieth centuries, the natural monopoly features of infrastructure industries, together with their strategic roles, have been important elements in state intervention. The aim of this paper is to evaluate what relative weight was attached to market failure problems on the one hand and geo-political factors on the other. For the period 1830–1939, how far were geo-political factors stronger than natural monopoly problems in accounting for the scale of intervention in the various countries of the Western World? How far did the policy instruments for security and market failure overlap? Whilst most of the infrastructure sectors are covered – including internal telecommunications, coal, gas, shipping, electricity and water – special attention is devoted to international submarine telegraph tables and railways. The paper concludes by demonstrating strong differences between Britain and USA on the one hand and Continental Europe plus Japan on the other.

Introduction

Studying institutions, and their changes over time, often requires a long perspective. Here the focus is Europe's infrastructure industries from roughly 1830 to 1939 – railways, submarine telegraph tables, internal telecommunications, coal, gas, shipping, electricity and water. The main concern is with the control and direction exercised by governments over these industries – 'regulation' in its widest sense or 'governance'. It has been argued that it is a mistake to attribute *differences* across countries (Europe, USA and Japan) in such business-state relationships simply to ideological stances (Millward 2008, 2011). For example Baily (1993) compared labour productivity in several sectors, including airlines, in Europe and America in the mid-twentieth century and concluded that the higher performance in the latter was due to a commitment to competition. In fact the early spread of 'national flag carriers' in Europe – Air France, Alitalia, Lufthansa, Iberia, KLM, BOAC etc. – was due to the perception of air space and airlines as key elements of national security, including links to the colonies. America was, geo-politically, very different from

Europe's hotchpotch of contiguous nation states with empires. The regulatory regime associated with IATA was intended to protect Europe's airlines against American competition. The same point applies in telecommunications to the contrast between arms' length regulation of AT&T in the nineteenth and twentieth centuries and the entrenchment of telephone and telegraph in government departments in Europe. Recently Bogart (2009) has recognised this in his study of nineteenth-century railways but not Wallsten (2005) in his study of telecommunications. Moreover some of the differences between service sectors and the changes over time in the form of governance have technological, fiscal and geo-political origins rather than being due to any strong shifts in ideological commitments, one way or another, to competition and capitalism. This is as true of the rise of municipal enterprise in the nineteenth century as to the privatisation of telecommunications in the late twentieth century, facilitated as the latter was by the new technologies of email, satellite and computer which broke the market stranglehold of the telephone monopolies. The loss of colonies and the emergence of jet aircraft undermined the national airline flag carriers. None of which is to deny the power of ideologies in some contexts to effect major changes (1980s Britain, Portugal 1960–90, perhaps late 1940s France) and their role as an ex post rationalisation of policies.

For each country there is however little doubt that through the nineteenth and twentieth centuries, the natural monopoly features of infrastructure industries have been an important element in state intervention. The aim of this paper is to evaluate what relative weight was attached to such market failure problems on the one hand and geo-political factors on the other. For the period 1830–1939, how far were geo-political factors stronger than natural monopoly problems in accounting for the scale of intervention in each (mainly Western) European country? How far did the policy instruments for security and market failure overlap? This is a big topic with many empirical implications. It has not to my knowledge been analysed in any depth for the nineteenth century so I hope my initial foray here will further research. In the absence of any obvious measure of the scale of intervention, I shall here consider the range of interventions between informal ministerial pressure, arms' length legislative regulation of prices and service levels, through various forms of government ownership to administration of the service within government departments, as indicative of the intensity of state intervention. It will become apparent that the position on the spectrum in which the economic organisation of a country's particular infrastructure industry was located depended on two factors: one was the importance of security and market failure problems in the sector. The other was how well endowed each country was with the resources in question and how dominant they were in the case of international markets (coal, telegraph cable, shipping).

Regulation for market failure

The behaviour of governments is examined here from the perspective of controlling natural monopolies and ensuring security of supply, and these two aims closely reflected the specific economic and technological characteristics of infrastructure industries. However, like other sectors of the economy, infrastructure industries were vehicles for other government objectives of wider economic and social significance, including the enhancement of the safety of products and services, working conditions, wages and the safety of the labour force, the government's fiscal base and the general health of the economy. In assessing the intensity of government intervention, we should also take into account the range of policy instruments used and these, for the nineteenth century, may be classed under four headings: first was the specification of property rights in terms of the institutional form of the sector (under company or civil law, for example), the ownership of shares and entry conditions. Legislation on prices, service levels, security of supply, wages, safety and working conditions was a second instrument but, thirdly, bureaucratic and ministerial power could be as defined in the legislation and also exercised as a discretionary power, by default. Finally there was the set of provisions for monitoring and reviewing performance, including supervisory agencies (like railway rates tribunals), the power of parliaments/Congress and their standing committees, consumer bodies, the judicial system.

Natural monopoly is usually defined as arising when the technology of the sector meant it was cheaper for one firm to supply the service than two or more. Nineteenth-century governments have often been viewed as wary of intervention and natural monopoly conditions need not have drawn in government. In nineteenth-century Europe they invariably did because of the problem of rights of way needed for network industries like telecommunications, railways, gas, electricity, trams and water supply. There were settings where space for development seemed unlimited – perhaps for American railways and, as we shall see, for the submarine telegraph cables developed by Britain, France, Germany and the USA. But for the large part, the European nation state governments were drawn to intervene in network industries because of the pressure to *regulate the use of land* and the compensation that should be paid for rights of way. Once rights were approved, a natural monopoly condition had been facilitated and governments could not then refrain from a) monitoring the soundness of companies, b) controlling profits, and c) adjudicating on the configuration of routes (cf. Millward 2005, Chapter 2).

The policy instruments in the nineteenth century had three features. First was the considerable variation across time and space. True there were some common elements. Legislation for prices and service levels could be found in many Western European countries. Also to be found widely was the encouragement of bidding processes for contracts, though there were many pressures fighting against this – the Pereire brothers in Paris in the 1840s protested that it would inhibit the attraction of capital for French railways – and corruption and bribery of politicians was common. In Spain there were low rates of return on foreign investment in companies running nineteenth-century railways but the financiers (Rothschilds etc.) often had lucrative deals in over-priced materials and rolling stock (Keefer, 1996; Radcliffe, 1976). Whilst state encouragement for contract bidding was common (promoting 'competition for the field'), there was much variation in other specifications of property rights. Allowing duplicate networks (promoting 'competition in the field') could be found in early Dutch and German railway development and in early telegraph, gas and water supplies in the UK (Chadwick 1859; Fremdling, 1999; Matthews, 1985) but it was not common and died out by the end the century. Allowing different operating companies to use the same track (promoting 'competition on the same line') was permitted in the 1838 Prussian railway legislation but this practice, now common, was absent in the nineteenth century.

A second feature was that there was no regulation of natural monopoly by the device of public ownership. Direct government involvement in the construction of networks or in operations was not a product of state concern over market failure. It was a massive intrusive form of intervention but was not used to control profits or adjudicate on routes. Indeed in many countries, public enterprises were treated similarly to private in their subjection to legislative ceilings on prices and profits. Thus the arms' length regulation by the nation state of small scale local gas, electricity and water utilities in France and the UK were the same whether the local enterprise was municipal or private. So also for Germany and Denmark where the regulation was often at a regional or municipal level. Arms' length regulation of prices could be found for the private railway companies in Britain as well as for state telegraph, rail and telephone lines in Scandinavia and Germany and for the concessionaires of the planned rail network in France (Bongaerts, 1983; Dunham, 1941; Hyldtoft, 1994).

Where the governance level was municipal, there is evidence that restraining monopoly prices was the last thing in the minds of some civic leaders. Several contemporary American commentators such as Meyer (1906) felt that the regulation of prices, concession periods, quality standards and environmental impacts was more restrictive in Europe than the USA for gas, electricity and tramways, and thereby detrimental to development – as also was municipal ownership. However, detailed studies of the incidence of municipalisation tell a different story. It would seem that municipal enterprise in gas, electricity and tramways flourished most in those towns which grew fastest in the nineteenth century. They had mounting public health and sanitation programmes which required financing and the profits of municipal enterprise proved to be a useful source of local tax relief. They were profit maximisers operating under the constraint of price ceilings set by regional or state governments and such price ceilings often became out-dated as technical progress lowered real costs in these sectors. Water supply was different. In contrast to the other utilities, this was an industry subject to diminishing returns (through the use of land). Faced with price ceilings and rising expectations about the quality of water supply, many private companies in the USA and Britain could not survive, and municipalisation followed (Dawson, 1916; Hyldtoft, 1994, 1995; Jacobsson, 2000; Millward, 1991, 2007).

Strategic aims and security of supply

This is a period when strong nation states were dominant. Some were new – Belgium, Italy, Germany, the Republic of Ireland. They were concerned to solidify the state both in the general sense of ensuring social and political unification and in the specific requirements of war and crises. Older nation states were under no less pressure. The case of the breakaway of S. Ireland in the 1920s from the UK is a good example. What did the new Irish Free State government do? It created a large number of state enterprises (well before anything comparable in Britain) in electricity, insurance and industrial credit, shipping, steel, aviation, dairies, sugar, and (later) surface transport. As part of the struggle for a 'self-sufficient and self-supporting state' said the Industry Minister, Lemass, the ailing sugar processing company was taken over by the state and preferences given to the Irish sugar beet producers. In 1936 the Turf Development Board was established to develop Ireland's key energy source. Historians have seen all this as owing little to socialist theory but rather to a desire for the new nation showing its firm control over these key strategic sectors and declaring it will produce a much more effective economy, than the ailing body it has left. Even by 2006 electricity, gas, railways and water were still public sector monopolies (Barrett, 2006, 2007; O'Grada, 1997).

We may start with a definition that resources and services are strategically important when they are perceived to be significant for national defence and unification and are not readily substitutable in the short run. The manner and intensity with which a country regulated them in the nineteenth century for the purpose of security of supply depended on i) how well endowed the country was, ii) how important the resource or service was geo-politically, iii) the technology and economics of the resource or service.

Each sector had its own strategic features in the nineteenth century. Thus telecommunications in its internal dimensions (overland telegraph and telephone links) and external (international overland and submarine telegraph) provided channels for information which were valuable not only commercially but also for diplomats, the navy and the military. National electricity grid networks were being developed in the 1920s and 1930s and were a key emerging energy source for industry. Like oil and coal, the security element arose in terms of supply, that is how well endowed each country was and far they could adjust to war and crises. Railways had immense importance for social and political unification as well as for transport in times of crisis. Well in advance of war, railway enterprises needed to know the military's plans for troop mobilisation and, as war approached, they had to have the capacity to mobilise supplies, and shift them to the war zones.

The aim here is to analyse in more depth the cases of the international telegraph and railways, in terms of the regulatory dimensions of market failure and strategic requirements. Domestic telecommunications were publicly owned by 1913 in most European countries, and in yet another significant contrast with the USA, the strategic importance of this sector was reflected in Japan by the rise of the state enterprise, Nippon Telegraph and Telephone Company (Anchordoguy, 2001). The development of electricity networks brought state intervention over the 1920–39 period because of the difficulty of overcoming the resistance of small scale private and municipal generators and distributors. The unequal spread of coal endowments across Europe had important effects in the pressures to develop hydro-electricity in France, Italy, Iberia and Scandinavia. The oil problem emerged only towards the end of our period but was enough to induce the British, French and German governments before World War I to take share ownership in companies exploiting Middle Eastern oil. These latter cases have been already analysed (Millward 2005, 2008). Here the focus is on the international telegraph and railways because of the significant insights they provide on the trade-offs between strategic and market failure issues in the period 1830–1939.

The international telegraph

By the 1930s, all European governments, including Britain's, were involved in share ownership, monitoring prices and making international agreements in the international telegraph industry, as well indeed to some extent in coal and shipping. These industries had been 'modern' for nearly a century; the period when significant mechanisation and/or electrification started in these industries was the mid-nineteenth century. The governments of Continental Europe had been intervening in the development of the international telegraph from the 1860s onwards. On the face of it, here lies a puzzle, in that, despite the strategic significance of the international telegraph and its monopolistic tendencies, Britain was different – its government intervened only loosely. The telegraph had three geo-politically significant features.

(1) It facilitated diplomatic links just as much as stimulating trade. Hoag's analysis (2006) of the trade effects of the telegraph suggests that in mid-nineteenth-century inland America, the average information lag fell from 10 days to zero.
(2) It was especially suitable for communications in numerical form, like commercial transactions, and hence of vital importance for international finance – a key element of Britain's economy and empire. News published in Europe could take three months to reach New Zealand even by the early nineteenth century, but by 1876 would arrive via cable in a matter of hours (O'Hara, 2010).
(3) It was readily adaptable as a major new form of communication for the army in times of war (as was the radio for ship to shore transmissions from the early twentieth century).

Whilst overland international cable routes followed quickly on the emergence of the electric telegraph in the 1840s, the submarine cable had to await the development of a good sea resistant insulator – gutta-percha, first used successfully in 1851. The main burst of successful activity in long distance submarine cables came in the 1860s. They were superior to overland routes which required rights of way from foreign governments so that messages were readily interceptible, despite the deliberations of the International Telegram Convention in Paris when governments agreed to ensure the non-interruption of international traffic (Silva & Diojo, 2006). The importance and vulnerability of the submarine cable was demonstrated most forcibly at the start of World War I when the allies quickly and easily cut the German Atlantic cables and linked the ends to their own systems. Prior to that, in the nineteenth century, both Germany and France were heavily dependent on the use of British owned cables (cf. later). Some indication of the strength of the British position comes from an estimate that in 1911 to isolate countries like Egypt and South Africa from telegraph connections would require five cable cuttings for each country. For the UK the figure was 49 (Kennedy, 1971, p. 741).

Long telecommunications lines were not necessarily natural monopolies. Research by McDougall (2006), Gabel (1994) and Lipartito (1988, 1989) suggest that the 'long line' for telegraph and especially telephone across America was not an obvious economic asset in the nineteenth and early twentieth centuries since regional markets required a lot of attention. The behaviour of AT&T was often predatory and it pushed through a monopoly position on the long telephone line partly by a public relations appeal about the unification of America. Canada provided an alternative scenario with linking agreements between regional groups. In the case of the submarine telegraph, the natural monopoly condition stemmed from the huge investment needed (and very much 'sunken') in charting a route, employing massive ships and winches to load, reel and lay the cable and in harnessing ships for the repair, rescue and maintenance needed through the life of the cable. Initially one cable to USA would have sufficed for Europe's commercial demands. Others might be competitive but essentially duplicates (abstracting that is, from national security considerations). These natural monopoly conditions did not stem from any need for governments to facilitate rights of way since the ocean bed was unused space. The only rights needed were for landing and here the USA was notable in its attempt to refuse landing rights in the USA to companies whose home government restricted access on home territory to indigenous enterprises (cf. France).

Although rising demand eventually fostered the commercial attraction of competitive lines, the above economic and technological characteristics of the telegraph did lead, indirectly, to another form of monopolisation, namely the dominance of the British companies and this stemmed from two features.

(1) Britain was initially the only country with the industry and capital to promote submarine cables (cf. Headrick, 1991, 2001). They required strong shipping support and shipping itself required coal. In the nineteenth century, as Table 1 shows, Britain was supremely dominant in both respects. As late as 1892 its coal output vastly exceeded the rest of Europe combined as did its steam merchant shipping fleet. Fremdling (1996) has provided us with a very clear picture of Britain's domination of the output and trade in coal with its exporting ships spreading all across Europe's Atlantic, North Sea and Baltic coastlines, reaching via river transport far inland to Paris, Berlin and beyond.

(2) The far flung British empire had many strategic outposts and land areas (Malta, Alexandria, Aden, Karachi, Ireland, Newfoundland, Cape Town, Durban) so it was much better placed for landing rights than were other countries. It also was careful to maintain good relations with Portugal since Lisbon, Cape Verde and the Azores were key staging posts for the Atlantic (Silva & Diego, 2006).

Table 1. Strategic instruments. Cables, shipping, rail track and coal. 1892.

	UK	Germany	France	Russia	Japan	USA
Coal output ('000 metric tons)						
	184,704	92,350[a]	26,179	6950	3176	162,686
Merchant shipping (net registered capacity in '000 metric tons)						
Sail	3080	986	407	323	n.a.	2172
Steam	4565	696	409	206	n.a.	2016
Total	7645	1682	816	529	102	2188
International telegraph cables (length in kilometres)						
Companies	155,814[b]	0	13,427	0	0	38,987
Governments	7804[b]	1541[b]	8432[b]	524	0	0
Total	163,618[b]	1541[b]	21,859[b]	524	0	38,987
Railway density and spread						
Track mileage open:						
Per 1000 sq. miles	197[c]	132	105	2.5	12	56
Per million population	527[c]	546	565	204[d]	48	2612

Notes: [a]Includes 20,978,000 metric tons of brown coal. [b]Includes all imperial links. [c]Britain, i.e. excluding all Ireland. [d]1897.

Sources: Telegraph cable data from Headrick (1991). Area of USA 1870–1950 (including lakes etc.) taken as 3,022,000 square miles as in US Bureau of the Census 1975. All other USA data from Mitchell (2003b). Other rail track, shipping and coal data from Mitchell (2003a, 2003c). Population and area data for Britain from Mitchell (1988). Area of Russia was 8,370,000 square miles according to Whitaker's Almanac for 1910 and this is used for 1892. Russian population from Mitchell (2003a). Other population data from Maddison (1991). Area of Japan taken to be 163,000 square miles as given in Whitaker's Almanac for 1910. Area data for France and Germany for 1892 relates to 1914 as in Whitaker's Almanac for 1938.

Hence Britain dominated the early cable market both in terms of the overland international cables and the submarine cables across the Atlantic, Mediterranean and round the Cape. The years 1865/6 saw the onset of the Atlantic (later Anglo-American) Telegraph Company's cable traffic from the UK to USA and the completion of a large overland route by a British/Indian company (the 'Indo-European Telegraph Department') from the UK to India via Germany and the Ottoman Empire. In 1868 the (British owned) Eastern Telegraph Company group developed a submarine route to India (and thence to the Far East) via the Bay of Biscay, the Mediterranean, Suez and the Red and Arabian seas. In 1870, a mainly overland cable was completed by a British owned company (the Indo-European Telegraph Company) through Germany, Russia, Armenia and Iraq to India. Routes round the Cape followed as did a Caribbean network and lines to Brazil, and by 1892 British-owned firms accounted for nearly two-thirds of the 221,000 kilometres of cables in the world (Headrick, 1991, p. 58) and 45% was the business of the Eastern Telegraph Company group. Of the 41 cablelaying ships in 1894, 28 were British (O'Hara, 2010). As Table 1 shows, even in the 1890s, the companies and governments of France, Germany, Russia and Japan together owned only 24,000 kilometres of international cables, that is, no more than 15% of the British total.

Despite the dominant part played by one firm and despite the strategic issues involved, the British government's regulation was extremely light. It did support British companies by providing marine surveys and sometimes guaranteed rates of return on invested capital, often in return for favourable rates for government traffic. It was not unlike the nineteenth-century British municipalities which encouraged the profit-making activities of their municipal enterprises to generate a revenue stream which could relieve local taxes (cf. above). The difference in cable was that the British national government did not get involved in ownership or operations in the nineteenth century. One exception was that the government, at the end of the century, despairing of private British companies crossing the Pacific, did discuss and eventually joined up in 1902 with the Canadian, Australian and New Zealand governments to set up a cross-Pacific link in the form of the Pacific Cable Board. Shortly afterwards, a new private company, the Commercial Pacific Cable Company, was formed which promised competition on the Pacific routes but even here the reality was rather different in that half the shares were owned by the Eastern Telegraph Company group which enforced high tariff rates so that its African-Asian cables were not strongly challenged.

So the British government intervened with a very light touch. This was not because there no market failures – the monopolistic elements were there, leading one Member of Parliament to declare he knew 'no monopoly in the world that is doing more injury to trade than the concentrated companies represented in the Eastern Telegraph Company' (O'Hara, 2010). Nor was the light touch due to the British government's ideological commitment to laissez-faire. There were security issues but the government had little to worry about because the industry (and the same was true of radio early on since Marconi had British nationality) was dominated by British firms whose cables could be readily commandeered in times of crises. As early as 1862 the government had set up a Joint Committee on Submarine Cables with representatives from the Atlantic Telegraph Company and the Board of Trade. For the time being informal links were enough.

The underlying issues were revealed once Britain lost its dominant position in the 1920s by which time German, French and American cable capacity had developed considerably (Table 2). The British government then promoted and subsidised the establishment of two companies. One was Cable and Wireless, a holding company with shares in the Eastern Telegraph Company group and also in Marconi. Secondly there was Imperial and International Communications Ltd., consisting of operations in:

- The Post Office radio and cable assets (not domestic);
- The Pacific Cable Board;
- The Eastern Telegraph system;
- The Marconi system.

Then in 1938 the Treasury acquired £2.6 million shares in Cable and Wireless. In 1944 there was an agreement with all Commonwealth countries that all telecommunications companies with international commitments should be publicly owned and in 1946 the Treasury acquired the remaining £27.4 million shares, followed in 1950 by the transfer of all Cable and Wireless assets to the Post Office (Laborie, 2006, p. 194; Lipartito, 2000, pp. 162–166; Chester, 1975, pp. 16, 110, 258, 453–456). To some extent, the same apparent volte-face occurred in the British government's attitude to coal. Faced with an industry whose productivity levels were looking very unfavourable relative to Germany and the USA, the state intervened on prices, and restructuring commenced in the 1930s (Greasley, 1990, 1995). During World War II, the industry was struggling to expand its output (in part because of old seams in an old industry) and after the war it was the first industry to be

Table 2. Strategic instruments. Cables, shipping, rail track and coal. 1923.

	UK[a]	Germany	France	Russia	Japan	USA
Coal output ('000 metric tons)				(USSR)		
	280,430	180,474[b]	38,566	12,700[c]	29,106[d]	596,841
Merchant shipping (net registered capacity in '000 metric tons)						
Sail	263	288	386	n.a.	n.a.	1254[e]
Steam	10,987	1546	1759	n.a.	n.a.	15,821[e]
Total	11,250	1834	2145	413[f]	3361[e]	17,075[e]
International telegraph cables (length in kilometres)						
Companies	251,592[g]	n.a[h]	33,399	0	0	136,146
Governments	46,210[g]	n.a[h]	31,544	2683	14,463	6475
Total	297,802[g]	n.a[h]	64,933	2683	14,463	142,621
Railway density and spread						
Track mileage open:						
Per 1000 sq. miles	222	207	123	6	61	83
Per million population	473	572	654	320[i]	172	2235

Notes: [a]Excludes S. Ireland. [b]Includes 118,249,000 metric tons of brown coal. [c]Includes 2,176,000 metric tons of brown coal. [d]Includes 157,000 metric tons of brown coal. [e]Gross tonnage. [f]1921. [g]Includes all imperial links. [h]German cables out of action, status undefined.

Sources: Telegraph cable data from Headrick (1991). Area of USA 1870–1950 (including lakes etc.) taken as 3,022,002 square miles as in US Bureau of the Census (1975). All other USA data from Mitchell (2003b). Other rail track, shipping and coal data from Mitchell (2003a, 2003b). Population and area data for Britain from Mitchell (1988). Russian population from Mitchell (2003a) and area for 1926 taken as the 1924 figure given by Lorimer (1946, p. 29). Other population data from Maddison (1991, Tables B-2 and B-3). Area for Japan taken to be 163,000 square miles as given in Whitaker's Almanac for 1910. Area data for France and Germany relate to 1924 as in Whitaker's Almanac for 1938.

nationalised. Of course industrial relations, poor working conditions and a strong trade union were intrinsic elements in this story but they would not have been able to play such a role in an industry less exposed to strategic needs.

Railways and the nation state

By the 1920s, regulation of the natural monopoly elements in railways had lead to a particular price structure and service pattern common to several Western European countries and with the following features.

(1) Behaving like classic discriminating monopolists, railway enterprises levied charges for particular consignments, passenger groups and routes which exceeded marginal costs by a surplus which reflected 'what the traffic would bear'. Hence the value categories for freight traffic and seating classes for passenger traffic. The evidence for France, Germany, Britain and Sweden is clear (Andersonn-Skog, 1999, pp. 248–249; Cain, 1973; Dougall, 1933; McMahon & Dittmar, Part III, June 1940, p. 191; Mierzejewski, 1996).
(2) Railway customers were given a 'universal service' in so far as marginal costs were small or ignored or could not be identified. Thus the rates and fares per kilometre within each class did not vary much within each country.
(3) Railway enterprises were open, because of their original monopolistic positions, to complaints, and hence vulnerable to pressures from governments, regional groups and business leaders for special marshalling facilities, insurance and other services, discounts, help for backward regions etc. (e.g. Anderson-Skog; 1996, 1999; Crompton, 1985, 1989; Mierzejewski, 1999, Vol. I, p. xv).

Market failure problems, therefore, lay at the heart of the railways' rate structure. However across Continental Europe in our period, the intensity of state intervention, the level and configuration of track development and the subsidies to capital cannot be explained simply as products of market failure problems. In the nineteenth century, there were two important features of national defence that affected railways, one internal and the other external. The promotion of internal social and political unification by trunk lines to all regions seemed to be a common aim of all governments, but the form of state intervention depended, as noted as a general point earlier, on how well advanced private rail development was.

Thus France developed a planned network operated by private companies in a concession system supported by government guarantees of rates of return since private capital tended to go to government bonds rather than industry. In Spain, partly due to inefficiencies with contracting, the subsidy was per kilometre of track laid (Cameron, 1961, p. 258; Keefer 1996; Laffut, 1983). In some countries, in their birth pains as nation states, there was a clear urgency attached to rail development. Direct state involvement in construction and operations emerged from the start in Belgium in the 1830s and the new state of Italy in the 1860s. In Sweden and Norway the sheer physical impediments and the initial failures lead impatient governments in the 1850s to construct the trunk lines (by the army in some cases) (Hedin 1967; Rawson, 1839; Soderland 1963). Britain, as the first mover, had completed its trunk network by the end of the 1840s. The private sector flourished so well that by the end of the century, the country was criss-crossed with rail lines. Its track mileage was not particularly large relative to population but it exceeded all others with respect to area covered (see Table 1).

Thus the form and intensity of intervention was very different in Continental Europe from that in Britain. This is reinforced when we look at external military threats. Continental Europe was a patchwork of nation states each with potential hostile states on their borders. It is clear from recent studies of nineteenth-century military requirements that a robust development of communications and energy resources was very important for military strength but ' . . . this would not have satisfied military requirements without political regulation' (Stevenson, 1999, p. 170). By the end of the century the mounting of a military operation involved three phases.

(1) Preparation. Communicating information about likely troop movements and energy requirements to governments and companies. Ensuring border areas were secure (rolling stock closely monitored).
(2) Mobilisation. This involved the calling up of reservists, making units mobile by assigning them to supplies and animals. This was not easy to plan. In 1912 it was estimated that 1000 train journeys would be needed in Britain.
(3) Concentration. Shifting units to zones of operation. For Moltke this meant that a nation's first concern must be 'to cover its territory with a network of railways which will ensure the most rapid possible concentration' (Stevenson, 1999, p. 174).

Clearly these phases required regulation of the railway systems such that the requisite track and rolling stock were available should war arise. In addition it required organisation and planning but, again, the intensity of state control of rail development varied with a country's geo-political position and the availability of resources. As an island economy with a huge navy and merchant fleet, British policymakers had perhaps less to worry about ensuring there were clear routes for rapid concentration. It appears that only in 1912–14 did the (Liberal) government (covertly) subsidise rail lines to Southampton which was the chief bottleneck (Stevenson, 1999, p. 174). It seems that direct ownership and operation of tracks by the state were less important than the development of good linkages between the railway companies and army planners. In France such links were forged via governments establishing committees. Some lines deemed to be unprofitable but important could be subsidised and, from 1880, no line could be closed without clearance by the Ministry of War (Doukas, 1945, p. 55). Subsidies were used, but only the Paris-Orleans line and then in 1906 the whole of the Western network was taken into public ownership. The 1870–1914 period did of course see tremendous growth in rail track mileage. For both France and Germany through-lines to the borders were vital. They rose from 9 to 13 in Germany in this period and 4 to 16 in France (Stevenson, 1999, p. 175). Germany was different in the scale and intensity of state intervention in railway development. The Prussian railways were nationalised in 1879 and the Prussian General Staff actually had a 'Railway Section'. As has been argued elsewhere (Millward, 2008, 2011), this particular form of organisation cannot be explained by issues such as the profits of the railways or the difficulties caused by the cheap rates offered to Russian grain shipments. It was an unambiguous sign of the desire to have a hands-on control of the system, readily comprehensible from Germany's geo-political position following the Franco Prussian war of 1870–71 and the growing threat from Russia.

Conclusions

In the period 1830–1939, the intensity of state intervention in infrastructure industries was stronger in Continental Europe than in Britain because the sectors in question had more geo-political significance for the former, constituting as it did, a set of contiguous and often hostile nation states. In Continental Europe the scale and nature of state intervention seems to have been determined more by geo-political factors than by problems of natural monopoly. The limited nature of the British government's involvement with infrastructure industries was a product of two factors. The first was that, as an island economy with a strong navy and merchant fleet, it had less need to worry about disruption to communications and trade nor to configure railways to meet war demands. It was also well blessed with railways as well as the shipping, coal and international telegraph industries where Britain was the first mover in international trade and sufficiently dominant for security issues to be dealt with on an informal basis rather than by outright acquisition and control, as in Germany, France and Scandinavia.

So far as local networks are concerned, strategic factors were largely absent and the main thrust of state intervention was on prices and service quality. If Germany and Britain are any guide, during the 1830–1939 period governments set ceilings on the prices of the small-scale gas, electricity and tramways enterprises (whether privately- or publicly-owned) but the municipalities were less interested in dampening down monopoly prices than in securing a share of monopoly profits. This tendency was particularly strong in fast growing industrial towns whose councils sought profits to relieve local taxes in the face of mounting public health programmes.

The British central government did not intervene strongly in cables (or coal) until the 1920s and 1930s when it had lost its dominant position in world supplies. Prior to that, before World War I, the British government seems to have taken the view that its endowment of the key strategic economic instruments (including railways), in conjunction with its geo-political position, meant its resources could be readily adapted to crises. Whilst then American and British government intervention 1830–1939 was limited largely to arm's length regulation of prices and service levels, this difference from Continental Europe was due to geo-political factors even though politicians and others often couched their policy stances in terms of an ideological commitment to laissez faire.

References

Anchordoguy, M. (2001). Nippon Telegraph and Telephone Company (NTT) and the building of a telecommunications industry in Japan. *Business History Review*, *75*, 507–541.

Andersson-Skog, L. (1996). From state railway housekeeping to railway economics: Swedish railway policy and economic transformation after 1920 in an institutional perspective. *Scandinavian Economic History Review*, *49*, 23–42.

Andersson-Skog, L. (1999). Political economy and institutional diffusion. The case of the Swedish railways and telecommunications up to 1950. In L. Anderson-Skog & O. Krantz (Eds.), *Institutions and the transport and communications industries* (pp. 245–266). Canton, MA: Science History Publications.

Baily, M. (1993). Competition, regulation and efficiency in service industries. *Brookings Papers in Microeconomics*, *2*, 71–159.

Barrett, S.D. (2006). Privatisation in Ireland. In M. Köthenbürger, H-W. Sinn, & J. Whalley (Eds.), *Privatisation experiences in the European Union* (pp. 265–288). Cambridge, MA: MIT Press.
Barrett, S.D. (2007). Transforming air transport in Ireland. In J. Clifton, D. Fuentes, & F. Comin (Eds.), *Transforming public enterprise in Europe and the Americas: Networks, integration and transnationalisation* (pp. 117–130). Basingstoke, UK: Palgrave.
Bogart, D. (2009). Nationalisations and the development of transport networks: Cross country evidence from railroad networks, 1860–1912. *Journal of Economic History*, *69*, 202–237.
Bongaerts, J.C. (1983). Financing railways in the German states 1840–60: A preliminary view. *Journal of European Economic History*, *14*, 331–345.
Cain, P.J. (1973). Private enterprise or public utility? Output, pricing and investment in English railways. *Journal of Transport History*, *1*, 9–28.
Cameron, R.E. (1961). *France and the economic development of Europe 1880–1914: Conquests of peace and seeds of war*. Princeton, NJ: Princeton University Press.
Chadwick, E. (1859). Results of different principles of legislation and administration in Europe: Of competition for the field, as compared with competition within the field, of service. *Journal of the Royal Statistical Society*, *22*, 381–420.
Chester, N. (1975). *The nationalisation of British industry 1945–51*. London: HMSO.
Crompton, G.W. (1985). Efficient and economical working? The performance of the railway companies 1923–33. *Business History*, *27*, 222–237.
Crompton, G.W. (1989). Squeezing the pulpless orange: Labour and capital on the railways in the inter-war period. *Business History*, *31*, 66–83.
Dawson, W.H. (1916). *Municipal life and local government in Germany*. London: Longmans, Green & Co.
Dougall, H.H. (1933). Railway rates and rate making in France since 1921. *Journal of Political Economy*, *35*, 289–333.
Doukas, K. (1945). *The French railroads and the state*. New York: Columbia Press.
Dunham, A.L. (1941). How the first French railways were planned. *Journal of Economic History*, *1*, 12–25.
Fremdling, R. (1996). Anglo-German rivalry in coal markets in France, the Netherlands and Germany 1850–1913. *Journal of European Economic History*, *25*, 446–599.
Fremdling, R. (1999). The Prussian and Dutch railway regulation in the 19th century. In L. Andersson-Skog & O. Krantz (Eds.), *Institutions and the transport and communications industries* (pp. 61–92). Canton, MA: Science History Publications.
Gabel, D. (1994). Competition in a network industry: The telephone industry 1894–1910. *Journal of Economic History*, *54*, 543–572.
Greasley, D. (1990). Fifty years of coal mining productivity: The record of the British coal industry before 1939. *Journal of Economic History*, *50*, 877–902.
Greasley, D. (1995). The coal industry: Images and realities on the road to nationalization. In R. Millward & J. Singleton (Eds.), *The political economy of nationalisation in Britain 1920–50* (pp. 37–64). Cambridge: Cambridge University Press.
Headrick, D.R. (1991). *The invisible weapon: Telecommunications and international politics 1951–1945*. New York: Oxford University Press.
Headrick, D.R., & Griset, P. (2001). Submarine telegraph cables: Business and politics. *Business History Review*, *75*, 543–578.
Hedin, L.-E. (1967). Some notes on the financing of the Swedish Railways 1860–1914. *Economy and History*, *10*, 3–37.
Hoag, C. (2006). The Atlantic cable and capital market information flows. *Journal of Economic History*, *66*, 342–353.
Hyldtoft, O. (1994). Modern theories of regulation: An old story: Danish gasworks in the 19th century. *Scandinavian Economic History Review*, *42*, 29–53.
Hyldtoft, O. (1995). Making gas: The establishment of the Nordic gas systems, 1800–1870. In A. Kaijser & M. Hedin (Eds.), *Nordic energy systems: Historical perspectives and current issues* (pp. 76–99). Canton, MA: Science History Publications.
Jacobson, C.D. (2000). *Ties that bind: Economic and political dilemmas of urban utility networks 1800–1990*. Pittsburg, PA: University of Pittsburg Press.

Keefer, P. (1996). Protection against a capricious state: French investment and Spanish railroads, 1845–1875. *Journal of Economic History, 56*, 170–192.
Kennedy, P.M. (1971). Imperial cable communications and strategy, 1870–1914. *The English Historical Review, 86*, 728–752.
Laborie, L. (2006). A missing link: Telecommunication networks and European integration. In E. van Leuten & A. Kaijser (Eds.), *Networking Europe: Transnational infrastructure and the shape of Europe 1850–2000* (pp. 187–215). Sagamore Beach, MA: Watson.
Laffut, M. (1983). Belgium. In P. O'Brien, *Railways and the economic development of Europe* (pp. 203–226). Oxford: Macmillan.
Lipartito, K. (1988). The telephone in the South: A comparative analysis, 1877–1920. *Journal of Economic History, 48*, 419–421.
Lipartito, K. (1989). System building on the margin: The problem of public choice in the telephone industry. *Journal of Economic History, 49*, 323–336.
Lipartito, K. (2000).Failure to communicate: British Telecommunications and the American Model. In J. Zeitlin & G. Herrigel (Eds.), *Americanisation and its limits: Reworking US technology and management in post-war Europe and Japan* (pp. 152–179). Oxford: Oxford University Press.
Lorimer, F. (1946). *The population of the Soviet Union: History and prospects*. Geneva: League of Nations.
MacMahon, A.W., & Dittmar, W.R. (1939/1940). Autonomous public enterprise – the German railways. *Political Science Quarterly*, Part I (LIV, Dec. 1939, 481–513), Part II (LV, March 1940, 25–52), Part III (LV, June, 176–198).
Maddison, A. (1991). *Dynamic forces in capitalist development*. Oxford: Oxford University Press.
Matthews, D. (1985). Rogues, speculators and competing monopolies: The early London gas companies, 1812–1860. *The London Journal, 2*, 39–50.
McDougall, R. (2006). Long lines: AT&T's long distance network as an organisational and political strategy. *Business History Review, 80*, 297–327.
Meyer, H.R. (1906). Municipal ownership in Germany. *Journal of Political Economy, 14*, 553–567.
Mierzejewski, A.C. (1996). The German National Railway Company confronts its competitors 1920–39. *Business and Economic History, 25*, 89–102.
Mierzejewski, A.C. (1999). *The most valuable asset of the Reich: A history of the German National Railway* (Vols 1 & 2). Chapel Hill: University of North Carolina Press.
Millward, R. (1991). The market behaviour of local utilities in pre-World War I Britain: The case of gas. *Economic History Review, 44*, 102–127.
Millward, R. (2005). *Private and public enterprise in Europe: Energy, telecommunications and transport c.1830–1990*. Cambridge: Cambridge University Press.
Millward, R. (2007). La distribution de l'eau dans les villes en Grande-Bretagne au XIXe et XXe siècles: Le gouvernement municipal et le dilemme des compagnies privées. *Histoire, Economie et Société, 26*, 111–128.
Millward, R. (2008). Business and the state. In G. Jones & J. Zeitlin (Eds.), *Oxford handbook of business history* (pp. 529–557). New York: Oxford University Press.
Millward, R. (2011). The institutional economic history of infrastructure industries c1830–1990: Ideology, technology, geo-politics? In M. Finger & R. Künneke (Eds.), *International handbook for the liberalisation of infrastructures* (pp. 87–99). Basingstoke: Edward Elgar.
Mitchell, B.R. (1988). *British historical statistics*. Cambridge: Cambridge University Press.
Mitchell, B.R. (2003a). *International historical statistics: Europe 1750–2000* (5th ed.). London: Macmillan.
Mitchell, B.R. (2003b). *International historical statistics: The Americas: 1750–2000* (5th ed.). London: Macmillan.
Mitchell, B.R. (2003c). *International historical statistics: Africa, Asia and Oceania 1750–2000* (4th ed.). New York: Stockton.

O'Grada, C. (1997). *A rocky road: The Irish economy since the 1920s*. Manchester: Manchester University Press.
O'Hara, G. (2010). *Britain and the sea since 1600*. London: Palgrave Macmillan.
Radcliffe, B. (1976). Railway imperialism: The example of the Pereires' Paris-St. Germain Company. *Business History*, *18*, 66–84.
Rawson, R.W. (1839). On railways in Belgium. *Journal of the Royal Statistical Society of London*, *2*, 47–62.
Silva, A.P., & Diego, M.P. (2006). From host and hostage: Portugal, Britain and the Atlantic cables. In E. van Leuten & A.Kaijser (Eds.), *Networking Europe: Transnational infrastructure and the shape of Europe 1850–2000* (pp. 51–60). Sagamore Beach, MA: Watson.
Soderlund, E.F. (1963). The placing of the first Swedish railway loan. *Scandinavian Economic History Review*, *11*, 43–59.
Stevenson, D. (1999). War by timetable? The railway race before 1914. *Past and Present*, *162*, 163–194.
Wallsten, S. (2005). Returning to Victorian competition, ownership and regulation: An empirical study of European telecommunications at the turn of the century. *Journal of Economic History*, *65*, 693–722.
Whitaker & Sons. (1910). *Whitaker's almanac*. London: Whitaker & Sons.
Whitaker & Sons. (1938). *Whitaker's almanac*. London: Whitaker & Sons.

Infrastructure and nation building: The regulation and financing of network transportation infrastructures in Spain (1720–2010)

Germà Bel

Department of Economic Policy, Universitat de Barcelona, Barcelona, Spain

This paper analyses Spanish infrastructure policy since the early 1700s: road building in the eighteenth century, railway creation and expansion in the nineteenth, motorway expansion in the twentieth, and high speed rail development in the twenty-first. The analysis reveals a long-term pattern, in which infrastructure policy in Spain has been driven not by the requirements of commerce and economic activity, but rather by the desire to centralise transportation around the country's political capital. As commerce has been unable to sustain the development of this policy, regulation and subsidies from the national budget have regularly been used to decide the priorities regarding infrastructure creation and to fund the development, maintenance, and operation of the networks.

When high roads, bridges, canals, etc. are in this manner made and supported by the commerce which is carried on by means of them, they can be made only where that commerce requires them, and consequently where it is proper to make them. Their expense too, their grandeur and magnificence, must be suited to what that commerce can afford to pay. They must be made consequently as it is proper to make them. A magnificent high road cannot be made through a desert country where there is little or no commerce, or merely because it happens to lead to the country villa of the intendant of the province, or to that of some great lord to whom the intendant finds it convenient to make his court.

Adam Smith, *The wealth of nations* (1776, vol. III.V.I, pp. 95–96)

Introduction

At the end of 2010, Spain has the largest High Speed Train network in Europe and the second largest in the world after China. Nonetheless, HST passengers in Spain represent only about 6% of those in Japan, and less than 20% of those in France. As far as airports are concerned, Spain is the only relatively large EU country with a large air market where ownership and management are strictly public and centralised, recalling communist models. Spain's motorway policy is singular as well: the financing model is hybrid, with 25% of the network tolled and 75% toll-free, with a sharp territorial asymmetry in the distribution of tolls throughout the country. These distinctive characteristics of Spain's transport infrastructures are the result of a long term policy that has used infrastructure for nation building rather than as a tool for creating a transportation system able to foster productivity and economic growth.

Centralisation of transport infrastructure, and particularly that of the railway systems, has received some attention in the literature, particularly in works on the history of transportation in Australia.[1] The centralisation of state railways in Australia has been extensively studied in Wotherspoon (1979), Docwra and Kolsen (1989), and more recently Gray (2009), who concludes that this centralisation emerged from the political desire to establish and maintain territorial control in the form of a hierarchical governmental and administrative system managed from capital cities. Other cases of centralisation, such as that of Argentina, seem to be more the result of the development of private industry, with no deliberate desire on the part of the central government to maintain control over it (Fleming, 1991).[2]

This paper studies whether the radial design of surface transportation networks around Spain's capital city, Madrid, reflects a deliberate desire of the central government. We analyse the infrastructure policy implemented in Spain since the early 1700s: road building in the eighteenth century, railway creation and expansion in the nineteenth, motorway expansion in the twentieth, and high speed rail development in the twenty-first. Our study contributes to the literature in several ways. First, the analysis is undertaken at nation-wide level. Second, we adopt a long-term perspective, which allows us to trace the centralisation pattern back to the period preceding the first railway extension. Third, and more importantly, we analyse network regulation and financing as the main tools used by government policy to centralise the surface infrastructure networks.

Our analysis reveals a long-term regularity: infrastructure policy in Spain has not been driven by the requirements of commerce and economic activity, but rather by the desire to centralise transportation around the political capital of the country. Because commerce could not sustain the development of such a policy, regulation and subsidies from the national budget have regularly been used to decide the priorities regarding infrastructure creation and to fund the development of the networks.

Summing up, this paper explains why national policies on regulation and financing of transportation infrastructure in Spain are unique among the developed countries. Starting in the eighteenth century, we see that transportation policy in Spain has been driven by a philosophy of nation building rather than by the needs of economic activity, and that centralisation has been the main driver of this policy.

Spain after the War of Succession: building a country like France, and a capital like Paris

Carlos II, the last king of the Hapsburg dynasty, died in 1700 and was succeeded by Philippe de Bourbon, Duke of Anjou and grandson of Louis XIV, the Sun King, who would reign in France until 1715. The new monarch, now known as Felipe V, arrived in Madrid in January 1701, and reorganised both the court and the government in accordance with French models. As Elliott (1963) points out, Spanish government offices now mimicked their French counterparts and leading governmental positions were awarded to Frenchmen brought to Madrid by the new king.

Shortly after the accession of Felipe V in 1702 the War of the Spanish Succession broke out, as the Austrian Empire fought to regain its right to the throne. This war began as an international conflict, but it also acquired an internal dimension, since the territories of Castile and Navarre aligned with the French party while the territories of Aragon supported the Austrians. The war, which 'took the form of a Bourbon conquest of Catalonia, Aragon and Valencia' (Ringrose, 1996, p. 139), came to an end in 1714/1715, with the consolidation of the new Bourbon dynasty.

During the course of the war, the first decree of the *Nueva Planta* was passed in June 1707, which abolished the institutions of the València and Aragon Kingdoms and reorganised the judiciary, introducing the ruling of Castilian courts. Soon after the end of the war, two decrees were published in November 1715 and January 1716, the first concerning the Balearic Islands and the second Catalonia, which reorganised the judiciary as well, and imposed the use of Castilian as the official language in administration. The monarchy also took control of the taxing system in the territories of the Aragon Crown. This move eased the pressure on the state's finances, together with the reduction of expenditure brought about by the loss of Italy and the peace with Holland after the Treaty of Utrecht in 1713. In the institutional area, 'local Courts were abolished, the Crown started to sell off the alderman positions in the municipal Councils, the courts joined the Castilian system and the mayors (*corregidor*) were appointed from Madrid. In this way, not only was royal authority strengthened in Castile, but Castilian practices were extended to the Crown of Aragon' (Ringrose, 1994, p. 296).

Spain was transformed *de iure* and *de facto* into a centralised state (Elliott, 1963). The centralist and 'Castilian' orientation of the new dynasty, which had already been evident since Felipe V's accession to the throne, was accentuated after the war. The new dynasty had come from France and adopted the French model as a reference in the planning of power. Like Paris in France, Madrid became the capital of a state in which the political power was monopolised by central institutions. From that moment on, the project of turning Spain into France, and Madrid into a capital like Paris, gathered pace. As explained in Álvarez Junco (2001, pp. 102–103):

> the idea was to imitate the French model... This meant reconstructing the Armada, centralising the administration, improving the roads... What was good for the monarchy was good for Spain. And what had strengthened the French monarchy would benefit the Spanish one as well.

The new plan met with little opposition; Castile's elite saw the decision to follow the French administrative and cultural model as a way to strengthen the political organisation of the monarchy and to redress the country's decline (Lluch, 1996, p. 122). Borja de Riquer (1994) notes that the political elite opted to unite and centralise as a way of establishing a nation. In short, Bourbon reformism was the adaptation of French reformism in Spain (Vicens Vives, 1952), and from that time onwards the French influence was omnipresent.

Road policy in eighteenth-century Spain: the birth of the radial state

The road network of Hapsburg Spain had its origins in the Roman network of the third and fourth centuries. It presented some of the basic characteristics of mesh connections and formed a decentralised network (Madrazo, 1984, pp. 27, 151–152), which was much less dense in the northwest of the peninsula (Galicia and Asturias) and in the southwest part of the central plateau or *Meseta* (Villuga, 1546). This network was characterised by itineraries from east to west that ran through the large river valleys, routes from north to south linked to the east-west ones, and some diagonal routes that also connected them to the roads above (Uriol, 1985, p. 556–557). In essence, this is the map of the roads inherited by the Bourbon dynasty at the beginning of the eighteenth century.

After the consolidation of the new dynasty's political power, the reform of the communications system was undertaken. The postal service, which had been contracted out to private interests since the sixteenth century, was reformed. In 1716, Felipe V completely internalised the service, which was to be provided directly by the Crown. Likewise, 'as part of the centralising policy of the Bourbons... who undertook the construction of a modern road network in Spain funded by the Treasury of the Crown' (Uriol, 1977, p. 837). The bases of this process were legally established in the *Reglamento General para la Dirección y Gobierno de los Oficios de Correo Mayor y Postas de España, en los viajes que se hicieren*, approved on 23 April 1720. The regulation introduced the *Relación de las Carreras de Postas* (List of the King's Roads), indicating the origin and destination of each journey. The list set up a radial network based on Madrid, going to (1) Bayonne, via Pamplona and Irun; (2) to Barcelona and the French border; (3) to València; (4) to Murcia and Cartagena; (5) to Cádiz via Seville; (6) to Badajoz; and (7) to Galicia via Medina del Campo (with a branch towards Salamanca). The network was completed by a parallel road along the Mediterranean between Barcelona and Alacant (with a link to Teruel), another parallel to the Portuguese border, between Benavente and Seville, and the connection between Burgos and Medina del Campo, via Valladolid. The King's Roads are shown in Figure 1.

The network was mainly designed to facilitate the dispatch of orders from the Court and the transmission of information to the capital, and it established a radial system that has remained unaltered until today.

Following the tradition established after the end of the Roman Empire, the building and maintenance of the motorways were the responsibility of the municipalities, which had to bear the costs of a process that was authorised by the Crown. This funding system turned out to be unable to achieve the objectives sought by the Crown for the King's Roads; the limited trade traffic did not generate sufficient revenues for municipalities to cover the costs of their investment. Therefore, in 1747 Fernando VI issued a *Real Cédula* (Royal Decree) which raised for the first time the possibility of building the King's Roads with funding from the Crown's Treasury (Uriol, 1978, p. 535). Thus, for the first time in the history of Spain, the state assumed direct financial responsibility for road construction.

The *Tratado legal y político de caminos públicos y posadas* (Fernández de Mesa, 1755) explicitly proposed that the cost of the creation of King's Roads should be covered by the Crown, and that the rest of the roads (though not the radial ones) should be financed by those who benefited from them, mainly the municipalities. Road building was extensively financed by the Treasury during the reign of Carlos III, who came to the throne in 1759. Indeed, in 1761 a general road plan was introduced, inspired by Bernardo Ward's recommendations written in 1760 (and later published in 1762). Ward's work defined six basic radial roads, coinciding

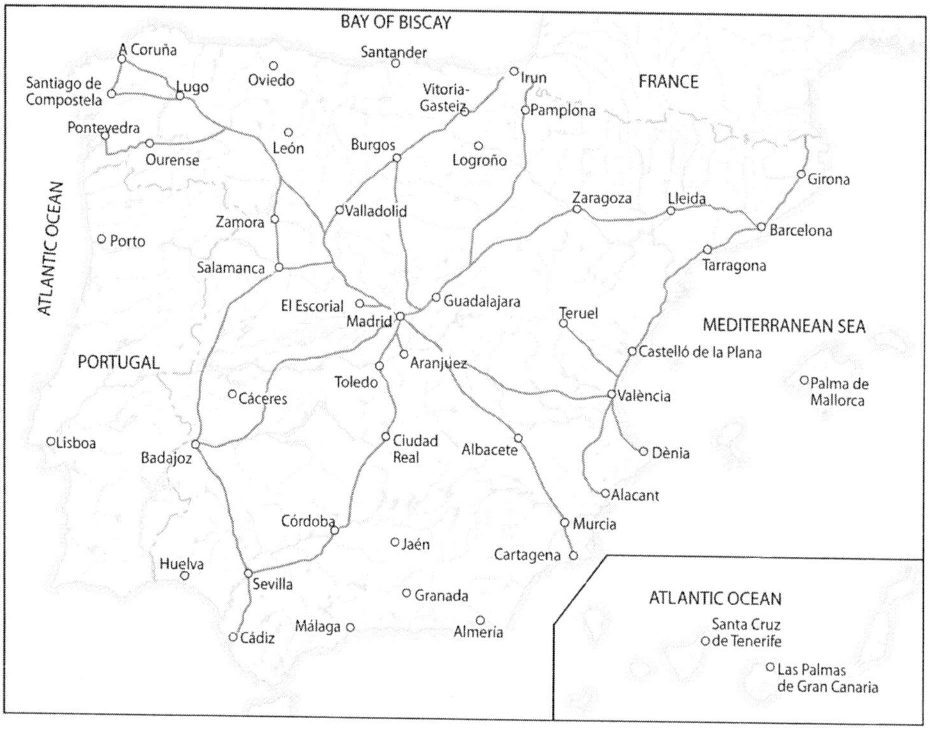

Figure 1. The King's Roads defined in 1720.

with the radial King's Roads in the 1720 Regulation: 'Spain needs six big roads, from Madrid to Corunna, to Badajoz, to Cádiz, to Alacant, and to the French border, through Bayonne and through Perpignan; and from these, other roads must go to several sea Ports and other main cities' (Ward, 1762, p. 55).

Ward urged that the building of these large roads should be funded by the Treasury. His proposals were included in the Decree issued on 10 June 1761, which included specific Treasury funding for the roads from Madrid to Andalusia, Catalonia, Galicia and València (Menéndez Pidal, 1951, p. 123). To finance these roads a national salt tax was instituted, initially for 20 years, then successively extended until 1801. This was the funding mechanism for the radial road network.

Thus, the map of the King's Roads instituted by Felipe V was consolidated as the radial road map of Spain by means of the 1761 decree, which established the six core axes with their six main roads. Gómez Mendoza (2001, p. 349) describes the reasons behind this design as follows: 'By means of a network of this type, the Bourbon legislators sought to satisfy administrative objectives (post and information), military, and of public order (the supply of goods to Madrid). They ignored purely economic objectives'. In fact, the question of efficiency and the contribution of the transport system to productivity and the economy were neglected in favour of administrative and political objectives (Madrazo, 1984, pp. 62–63, 159). Political and government interests were the priority, and a model was born that would become permanent in Spain.[3]

The establishment and development of the railway in Spain

Early days

After the failure of the initial rail projects in 1820 and the 1830s, the first successful railway concession was the one awarded in 1843 for a line of 28.5 km between Barcelona and Mataró, promoted by private interests that expected freight traffic to provide the major part of the demand (at this time the possibility of transporting passengers was barely considered). Shortly afterwards, in 1845, the Marquis of Salamanca was awarded a line between Madrid and Aranjuez, covering 49 km. Work on this line started in 1846, but was interrupted at the end of 1847 due to financial difficulties. In contrast, the Barcelona-Mataró line, which had received no public support, was completed and opened on 28 October 28 1848, becoming the first operating railway in Spain.[4] Work on the Madrid-Aranjuez line recommenced after obtaining funding from the Treasury, and the line was finally inaugurated on 9 February 1851.

Other stretches of railway were built in those early years. The Xàtiva-València line, covering 57 km, opened in 1854; the mining line Sama de Langreo-Gijón, was completed in 1856. In Catalonia the railway expanded rapidly (Pascual Domènech, 1999, pp. 91–94), with lines linking Barcelona with Granollers and Molins de Rei (and its extension to Martorell), and the stretch between Sabadell and Montcada which linked up with the Barcelona-Granollers line. Most of these first lines reflected a real interest in exploiting the existing road traffic flows (Comín Comín et al., 1998, p. 41), with the exception of certain lines, such as the one from Madrid to Aranjuez and then to Albacete, which had been financially guaranteed by the state, thus violating the current legislation.

The genesis of the radial railway in Spain: the Railways Parliamentary Committee of 1850

In the mid-nineteenth century, government involvement in railway expansion increased. After an initial failed attempt by the Minister Bravo Murillo in 1847 to pass a General Railway Act, the project was taken up again in 1850 by the Minister Seijas Lozano. His plan was appraised that year by a Congressional Committee (the Olózaga Committee, named after its chairman). The fundamental objective of the planned network was to link the political capital with the maritime periphery, according to Mateo del Peral (1978, p. 56), because 'The members of the commission, who resided in the capital, saw secondary railways arising with partial routes, which aimed to link close centres of production, normally within the same provincial area, while Madrid remained relatively isolated and in danger of accentuating its isolation.'

Among all the points considered by the commission, we are particularly interested in the decisions on the shape of the network and the choice of the priority routes. In agreement with the government, the commission tried to establish and subsidise a number of main routes at the public treasury's expense: (1) from Madrid to Bayonne, with branches to the major ports of the north and northeast, and one from Valladolid in the direction of Porto, (2) from Madrid to Cartagena, with branches to Alacant and València, (3) from Madrid to Barcelona via Zaragoza, (4) from Madrid to Lisbon through Extremadura, and (5) from Madrid to Cádiz as a continuation of the Lisbon line or a branch of the one to Cartagena (Congreso de los Diputados, 1850, p. 161).

The testimony of prominent military men, engineers and financiers provide valuable information regarding the criteria adopted for the railway development.[5] For example, General Laureano Sanz emphasised the importance of supplying the capital: 'the lines that I consider best for the prosperity of the country are those that connect the capital of the monarchy with the Ocean and the Mediterranean; i.e. the ones that can bring anything necessary to the centre of the monarchy' (Congreso de los Diputados, p. 164). Mr Mazarredo also agreed that railway lines should start from Madrid because:

> The situation of the capital of the Monarchy in Madrid places the city at a certain disadvantage as the area and the surroundings are lacking many of the elements that allow capitals great development in other nations. Madrid is not a producer, but it finds itself in possession of the capital status; and without great public disruption this cannot be changed... Thus, the rest of Spain should compensate the advantage of not being subjugated by Madrid. Madrid, as a capital, has great needs; and these are not covered by its own production; so it is necessary that the producer should meet these needs in order to supply to this point the foodstuffs and the basic products which are consumed in so much abundance. (Congreso de los Diputados, 1850, p. 165)

Another influential soldier, Brigadier Buenaga, completed the argument in favour of a radial network stating that 'the Madrid of 1808 was not the Madrid of 1850; since that time, many of the provincial antipathies have been calmed, power has been centralised, political constitutions have been transformed and have given new fuel and vigour to the State, the Court has visibly gained in importance, and it is and will be, for a long time at least, the strategic political point' (Congreso de los Diputados, 1850, p. 190). The justification of the grounds for a radial design with Madrid in the centre had been established. The issue was resolved by the commission, with just one main change: the elimination of the line from Madrid to Barcelona via Zaragoza, which, as far as the supply of the capital was concerned, was considered to be of minor interest.

Railway development and legislation: the Acts of 1855, 1870 and 1877

The first Railways Act, 1855, took as its basis the priority lines of the 1850 Commission, recovering the Madrid-Barcelona line via Zaragoza, and adding the extension to Málaga of the Cádiz line, the extension to the Galician coast of the line to Valladolid and the connection between Zaragoza and the Madrid-Irun line. These lines are shown in Figure 2. This is how the radial design of the railway system was established, since these were the priority lines which would benefit from state promotion and subsidies.

The railway policy adopted in Spain differed significantly from other European experiences where commercial interests were the driving factor of the railway network extension, such as the United Kingdom (Hawke & Higgins, 1983), Belgium (Derry & Williams, 1960), Germany (Fremlindg, 1983), and Italy (Fenoaltea, 1983). In some aspects, the Spanish policy resembled the policy carried out in France, especially in terms of the choice of the model of concession to the private sector with strong support from the public treasury. The main lines would be built at the state's expense, i.e. with guarantee of full recovery of investments and profitability. But while in 1838 the French parliament had rejected a plan to create a radial network emanating from Paris (Caron, 1983, p. 29), the Spanish Congress approved the

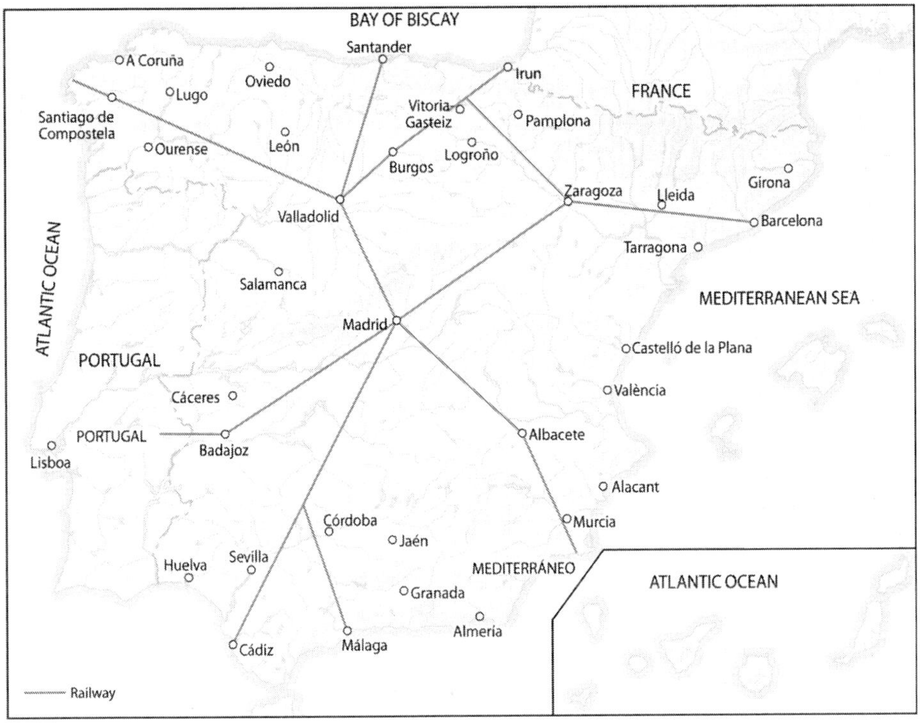

Figure 2. Railway priority lines approved by the Spanish Parliament in 1855.

construction under the state's guarantee of a radial network with its centre in Madrid.[6]

With the passing of the 1855 Act, the railway expansion gathered momentum apace. Between 1855 and the end of the 1860s almost 6300 km were granted; together with the 1100 km granted prior to 1855, the total figure amounted to 7363 km. However, not all the lines that were granted were actually built.

In 1864, the preliminary report for a new General Railway Plan, drawn up by a committee of engineers appointed by the Advisory Board of Civil Engineering, proposed as a first objective 'To link, in the best way possible, and taking advantage of the existing lines, all the capitals of province with the capital of the Monarchy' (Comisión de Ingenieros, 1864, p. 200). The report distinguished between main lines, which could receive financial help from the state, and secondary lines, which would only be granted if there were entrepreneurs willing to undertake their construction without state support. The main lines included all those necessary to complete the connection of the province capitals with Madrid, plus a few non-radial exceptions. The vast majority of transversal connections were proposed as secondary lines (Comisión de Ingenieros, 1864, pp. 201–204, 208–214).

A year later, in 1865, a special committee was appointed to assess the effects of the 1855 Act. The building activity induced by public subsidies had been such that the committee judged that the network was far larger than a country with Spain's population, wealth and economic activity actually needed, and that it was unsustainable with the resources available (Mateo del Peral, 1978, p. 101). Figure 3 shows the entirety of the network in late 1865.

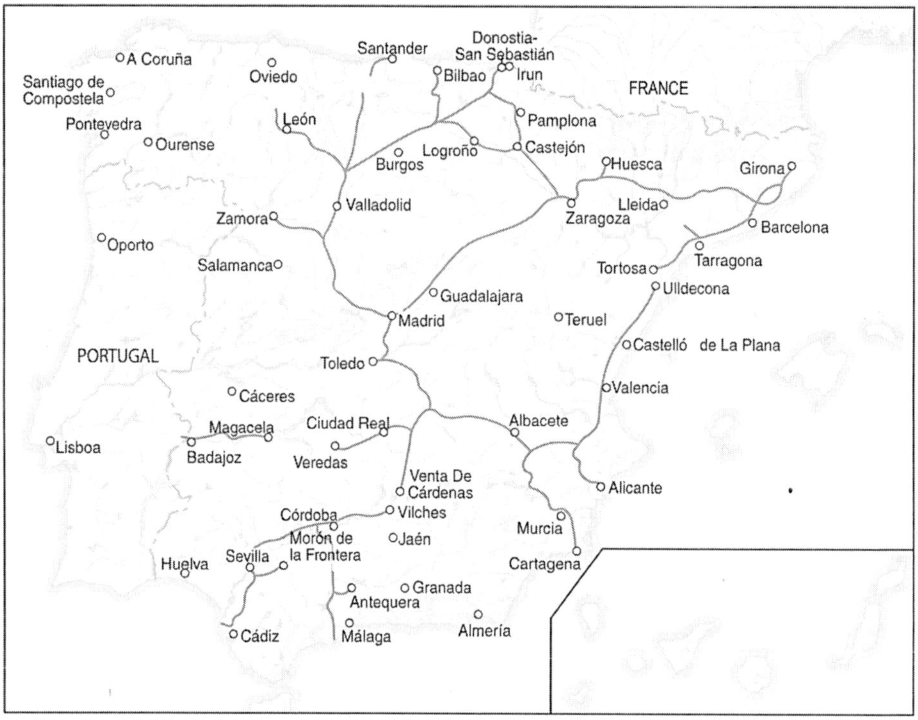

Figure 3. Railway network by the end of 1865.

The special committee drafted a proposal, completed in 1867, with a double objective. From the perspective of administrative and domestic policy, the proposal aimed to link Madrid to all provincial capitals not yet connected. As far as commerce was concerned, it sought to increase traffic by linking productive and trading areas which had previously been isolated; cross lines carried out their specific function of activating trade between the main export and import areas. Nevertheless, amid growing political and economic instability Isabel II was deposed in 1868, in the wake of what came to be known as the September Revolution.

The new government, though generally more liberal, took a keen interest in the field of public works, especially railways, and in 1870 drafted a bill proposing 2069 km of new lines. The primary objective was to connect Madrid with all the remaining provincial capitals not covered in the earlier plans (Teruel, Soria, Caceres, Jaén, Almeria, Salamanca, and Pontevedra) or whose licenscs were at the risk of expiring before they had even been implemented (Segovia, Cuenca, Huesca, and Huelva). These projects amounted to a total of 1284 km, most of which was to be built in eight years. In addition, another 800 km of mainly transversal lines were planned, to be built within 20 to 30 years. The government bill became law in 1870. Shortly after the end of the *Sexenio Revolucionario* and the return of the monarchy, in 1874, a new Act was passed in 1877. It would remain in effect for a long time; according to Mateo del Peral (1979, p. 130) the Act represented 'fidelity to the radial pattern . . . with the formulation of lines from Madrid . . . It did not solve the problem of the excessive centrality of the network.'

In short, railway regulation established the radial, centralised design of the Spanish network. In order to make this possible, the large-scale involvement of the state was needed in the awarding of grants and subsidies. These grants were primarily applied to the radial lines, the ones classified as first order or general service. Financial support from the state was crucial to the success of the construction process, and subsidies became the main fuel of the great investment in railways between 1855 and 1865, one of the most intense periods of investment in public works in Spain (Alzola y Minondo, 1899, p. 372).[7] The need for subsidies intensified with the construction of lines from the 1870s onwards (Herranz, 2008). In general, state subsidies were essential to support the planned network; its radial, centralised design could not be justified on the grounds of economic or commercial activity.

The establishment and development of motorways in Spain

The first phase of the (toll) motorways (1960s and 1970s)

Spain entered the 1960s without intercity motorways. A World Bank report of 1962 recommended the construction of a new motorway along the Mediterranean coast, from France to Murcia. This motorway would cross the major industrial, agricultural and tourist areas, the areas with the highest traffic intensity in the country. The report added that increasing traffic would argue in favour of the construction of fast roads in other regions in the near future. Five years after the World Bank Report, in 1967, the *Programa de autopistas nacionales Españolas* (PANE) was drawn up, which foresaw the construction of 3160 km of toll motorways. In 1972 the first sections were awarded: La Jonquera-Barcelona-Tarragona, and Montgat-Mataró-Bilbao-Behovia. Other concessions followed, such as Villalba-Villacastín-Adanero, Sevilla-Cádiz and Salou-València-Alacant. The prospect of toll motorways raised great expectations all over the country; the PANE update, the *Avance del plan nacional de autopistas* of 1972, planned 6430 km of toll motorways.

In Spain, the funding formula chosen in the 1960s was based on tolls, above all because of the insufficiencies of the country's tax system (Bel, 1999, pp. 124–126). In spite of the initial enthusiasm, however, the direct toll system did not encourage the expansion of the network, and the number of toll motorways in operation advanced very slowly. Of the 6430 km planned in 1972 and the 2042 km awarded in 1975, only 1807 km were working in 1985. Motorway concessions were abruptly terminated during the economic crisis of the 1970s. Figure 4 shows the stretches of toll roads operating in the territory.

It is easy to explain why toll roads were built where they were built. Funding by the user provides incentives to prioritise the routes with more business and growth potential since it offers better prospects for recovering costs. Naturally, the first significant concessions (1967–1968) were in the corridor with the greatest economic growth and, therefore, the greatest increases in traffic (particularly in goods) and with the highest expectations of profitability. In Spain, these were the stretches between La Jonquera and Tarragona, and between Montgat and Mataró, that is to say, in the corridor recommended by the World Bank in its report of 1962.[8] In late 1973, the only routes that had been awarded almost entirely were La Jonquera-Alacant and El Vendrell-Bilbao. Indeed, the Mediterranean and the Ebro Valley were the corridors in which economic activity and heavy traffic were growing the fastest, and together they accounted for more than two thirds of the total km of toll

Figure 4. Toll motorways built between 1967 and 1985.

motorways. Since they were the most dynamic corridors in terms of both passenger and goods traffic, they offered greater potential for obtaining toll revenues.

From tolls to budged funding: free motorways in the second phase (1980s–)

From the mid-1970s onwards, a series of problems that had emerged in the Spanish toll model began to be debated. Besides the very limited expansion achieved, another point of dissatisfaction with the policy of toll road concessions was the detriment to public finances in the medium and long term.

On coming to power in late 1982, the Spanish Socialist Party opted for a model of budgetary financing for the new motorways in their General Road Plan (PGC) 1984–1991.[9] There were several reasons for this change of policy. The two most important were the fact that the tax burden in Spain was much higher than in the 1960s – and so were the available funds from the budget – and the fact that the country's entry into the EEC and access to European regional policy funds provided additional budgetary resources for the construction of motorways.

The investment budget funding speeded up the provisioning process for motorways, and the PGC was completed in 1993. The result was a spectacular extension of the motorway network in Spain. By the end of 1993, 3045 km of PGC-toll-free motorways had entered into service.[10] Figure 5 shows the position of the new free motorway stretches in the Spanish territory. Significantly, 76% of the km corresponded to radial motorways, while only 24% were located in the transversal corridors.[11]

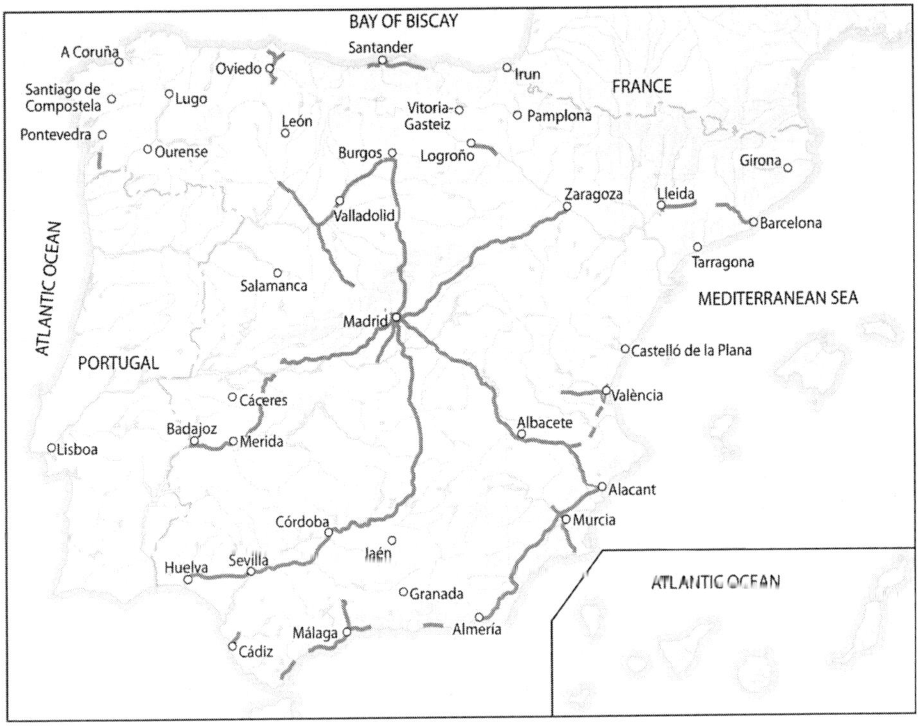

Figure 5. Toll-free motorways opened between 1985 and 1993.

The effect of the construction of PGC motorways was clearly described by the General Secretary for Surface Transport Infrastructures (Spanish Government), Emilio Pérez Touriño (1994, pp. 14–15): 'It is a network that has provided the radiality of our country with high capacity roads... as for the funding of high capacity, the network is focused on completing the radial connections of our country'. While some of the radial motorways were among the roads with most intense traffic, traffic in other radial motorways (i.e. Madrid-Badajoz and Burgos-Tordesillas) was less intense than that in many transversal routes that were not converted into free motorways (Bel, 2010, pp. 151–154).

In short, motorway construction began in the 1960s with the almost exclusive introduction of toll roads in the corridors with higher traffic and toll revenue expectations. In the second stage, the state took on the main role and used the public budget to turn the motorway network into a radial one, prioritising the radial roads that converge on Madrid. This policy diverged from that in the main European countries; the toll-free motorway extension in Germany was driven by commercial interests from the very beginning (Rothengatter, 2005); other important countries have consistently and continuously used tolls to fund the motorways network all over their territory, as in France (Fayard, Gaeta, & Quinet, 2005) and Italy (Greco & Ragazzi, 2005). In the Spanish case, the parallelism with the financial pattern followed in the railway extension is very striking. Moreover, it clearly illustrates the existence of an underlying pattern in the country's infrastructure policy, sustained over time, which responds to reasons that have not much to do with transport policy in a strict sense.

Railway modernisation from the 'kilometre zero': implementation and development of the Alta Velocidad Española (AVE)

The Spanish High Speed Train (AVE, not compatible with freight traffic so far) has become the rising star of the Spanish infrastructure policy. The investment in AVE lines shows the degree of primacy it has enjoyed, with enormous amounts of material resources and technical and human energy being devoted to its extension. According to the Ministry of Public Works' Strategic Plan for Transport Infrastructures, this trend is likely to continue until 2020.

The first AVE line became operational in 1992 between Madrid and Seville, and allows the non-stop journey between the two cities in two hours and 20 minutes. Spain decided to build a segregated high-speed network, as had been done in Japan and France (though in these countries conventional lines are compatible since they share the same gauge), and unlike Germany and Italy. Also unlike other countries, Spain decided to buy rail technology abroad rather than develop its own technology (Vickerman, 1997).

In 1995, the construction of the AVE Madrid-Barcelona-French Border line began, which started to run in 2003 between Madrid and Lleida and reached Tarragona in late 2006. At the end of 2007 the AVE started its services from Madrid to Segovia-Valladolid, and Málaga, in this last case from the line between Madrid and Seville. A branch was also opened from this line to Toledo in 2005. The AVE arrived in Barcelona in February 2008, and almost the entire stretch to the French border is currently in construction. In December 2010, the AVE between Madrid and València has become operational. Spain is now second only to China in terms of operative high-speed kilometres. See extension of AVE in Spain in Figure 6.

In Spain, the development of the AVE has followed political criteria and has not paid attention to the social return on this investment. The cost-benefit analysis for the Madrid-Seville line showed that this relationship could not be economically justified given its net negative benefits, and relatively low traffic, despite its large modal share in the corridor (De Rus & Inglada, 1997). A recent analysis of the profitability of the Madrid-Barcelona line showed that potential demand would not offset the large-scale investment, and that time savings would be small in aggregate terms, although it would manage to capture a large share of traffic (De Rus & Román, 2006).

Indeed, the policy of AVE extension has involved very high investment costs. Overall, cumulative investment until 2010 is close to €45 billion (Albalate & Bel, 2011), suggesting very low (or negative) rates of return. The total volume of traffic is very low, because the population served is relatively small and the competition from air transport is higher than in other countries, given Spain's geography and urban structure.

Countries that were pioneers in the development of high-speed networks in Europe (France and Germany) have granted priority to issues such as the potential traffic demand of the lines to modernise, the freight traffic needs, and so on – that is to say, to the core issues of transport policy. In this way, high speed rail extension in France has been restricted to lines generating a positive net social return of investments (Dunn & Perl, 1994; Vickerman, 1997). In Germany, the main consideration when designing the new lines (mixed passenger-freight) has been serving the highly profitable overnight traffic between the North Sea ports and the industrial areas and consumer markets in Southern Germany (Heinisch, 1992).

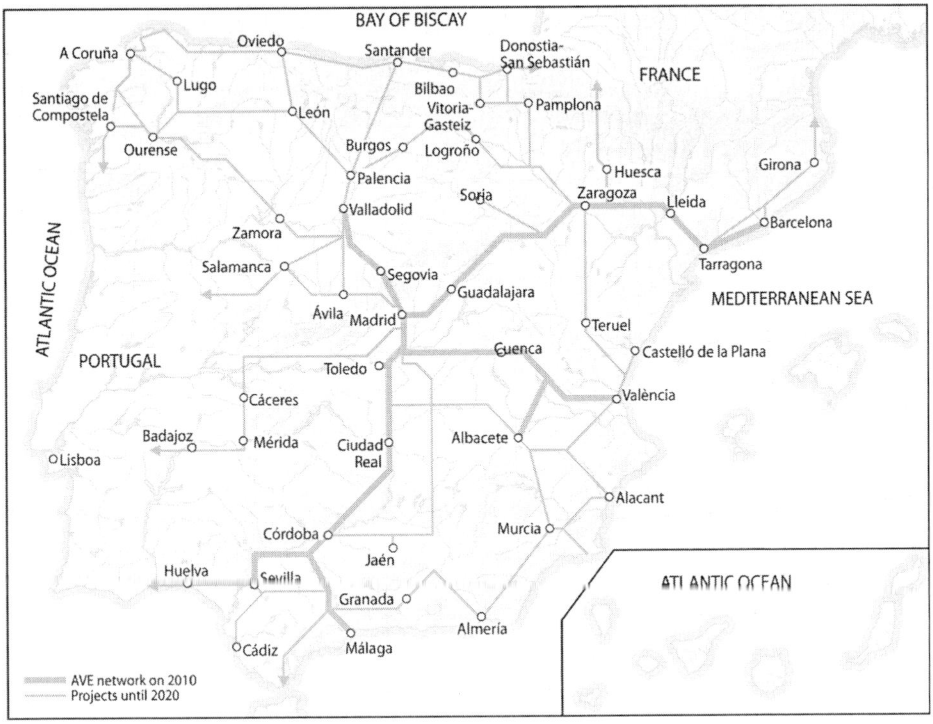

Figure 6. AVE network in Spain.

The case of Spain is different. The rationale for the extension of the AVE network was expressly declared on 25 April 2000 by the then Prime Minister José María Aznar, who declared the lynchpin of its infrastructure policy 'a high-speed rail network that, in ten years, will place all provincial capitals within four hours from the centre of the peninsula' (Aznar, 2000, p. 29) – whatever the cost, and regardless of the fundamental function that transport fulfils. So Spain is the country with the most high speed train kilometres in Europe, although its demand for services is very small compared with the pioneers: AVE passengers in Spain account for only about 6% of those in Japan, less than 20% of those in France, and about 30% of those in Germany.

Therefore, the development of the AVE, with its radial expansion, has followed a familiar pattern: the use of the law (regulation) and budget subsidies to achieve something that cannot be sustained by economic dynamics, that is, the radialisation of Spain's communication networks. It is yet another example of an infrastructure policy that has been in operation since the eighteenth century, more precisely since the enactment, on 23 April 1720, of the *Reglamento General para la Dirección y Gobierno de los Oficios de Correo Mayor y Postas de España, en los viajes que se hicieren*: the constituent act of the radial state.

Conclusion

The objective of this paper has been to analyse whether the radial design of surface transportation networks in Spain is due to centralist policies intended to establish territorial control and to maintain a hierarchical government administrative system managed from the capital city, Madrid. To do so, we have explored the infrastructure policy regarding roads and railways implemented in Spain between 1720 and 2010. Special emphasis has been placed on the analysis of network regulation and financing used by the national government in order to centralise the road and railway networks.

Our analysis draws attention to a long-term pattern: transportation infrastructure policy in Spain has been driven not by the requirements of commerce and economic activity, but by the desire to centralise the transport system around the country's political capital. Regulation and subsidies from the national budget have regularly been used to decide the priorities, because the development of this policy was not sustainable by commerce alone.

When designing their infrastructure policy, most developed countries have given priority to the requirements of the economic system. Instead, Spain has used that policy for the purposes of nation-building. The contribution of the transport system to productivity and efficiency has been systematically disregarded as a driver for infrastructure policy. The territorial asymmetries in toll versus non-toll motorways and overinvestment in national networks such as High Speed Rail can be explained within the framework of a philosophy that goes back three hundred years.

Acknowledgements

This research has received financial support from the Spanish Ministry of Science and Innovation (ECO2009–06946/ECON), from the Autonomous Government of Catalonia (SGR2009–1066), and from ICREA Academia. I have benefited from comments by Daniel Albalate, Joan Calzada, Judith Clifton, Xavier Fageda, Alfonso Herranz, Pedro Pablo Ortúñez, Ramon Perelló, and Emili Rosales. Comments received through the editorial process have been extremely helpful.

Notes

1. The effects of the radial railway network in the nineteenth century in Spain have been discussed in Nadal (1975) and Casañas Vallés (1977). However, they do not specifically discuss the issue of centralisation.
2. The early development of telegraphic communication systems was primarily driven by military purposes in most countries (Flichy, 1995, p. 19). France was an outstanding example of this trend, which resulted in a highly centralised network. However, the initially centralised design of the French network was soon seen as inefficient, and from the early 1830s on junction lines were built in the French telegraph network, so that a 'multi-connection' network could be created (Flichy, 1995, p. 30).
3. Note as well that on 16 January 1769 a Royal Order was issued establishing that distances in each road would be counted from Madrid. This constitutes the historical origin of the 'kilometer zero', which still exists in Spain.
4. It is worth noting that the first operating railway in Spanish territories was that between La Habana and Güines, in Cuba, which began service in 1837 (de Diego García, 1983).
5. The opinions expressed and recommendations made are available in *Información parlamentaria hecha por la Comisión de Ferro-carriles nombrada por el Congreso de los Diputados en 10 de enero de* 1850.
6. The way in which radial and transversal lines were combined in the development of the French network can be seen in Dunham (1941).
7. Tortella Casares (1973, pp. 168–173) estimates that public aid represented 49% of the capital paid out by railway companies until 1867. Researchers believe that it is impossible to evaluate all the subsidies awarded by the state (Comín Comín et al., 1998, p. 93).
8. Indeed, these first concessions have been highly profitable (Albalate & Bel, 2009).
9. See Boletín Oficial de la Cortes Generales (Congreso de los Diputados), Serie E, n. 160, 1 March 1986, pp. 1853–1882 (p. 1873).

10. In the same period, well under a hundred kilometres of new toll motorways were franchised by the state.
11. Detailed information on the routes and years in which stretches of free motorways were opened can be found in Bel (1994).

Printed primary sources

Olozaga's Commission (Spanish Congress, 1850). *Minutes of testimonies made in the Commission.*

Spanish Congress Gazette.

Spanish Official Gazette.

References

Albalate, D., & Bel, G. (2009). Regulating concessions of toll motorways: An empirical study on fixed vs. variable term contracts. *Transportation Research-A, 43*, 219–229.

Albalate, D., & Bel, G. (2011). Cuando la economıa no importa: Auge y esplendor de la alta velocidad en Espana. *Revista de Economía Aplicada, 55*, 171–190.

Alvarez Junco, J. (2001). *Mater Dolorosa. La idea de España en el siglo XIX.* Madrid: Taurus.

Alzola y Minondo, P. (1899). *Las obras públicas en España: Estudio histórico.* Bilbao: Biblioteca de la Revista de Obras Publicas.

Aznar, J.M. (2000, April 25). *Debate sobre la investidura del candidato a la Presidencia del Gobierno.* Diario de Sesiones del Congreso de los Diputados. Pleno y Diputacion Permanente. Ano 2000, no. 2.

Banco Mundial. (1962). *Informe del Banco Mundial. El desarrollo económico de España.* Madrid: BIRF.

Bel, G. (1994). *La demanda de transporte en España: Competencia intermodal sobre el ferrocarril interurbano.* Madrid: IETC.

Bel, G. (1999). Financiacion de infraestructuras viarias: La economıa polıtica de los peajes. *Papeles de Economía Española, 82*, 123–139.

Bel, G. (2010). *España, capital París. Origen y apoteosis del Estado radial.* Barcelona: Destino.

Caron, F. (1983). France. In Patrick O'Brien, *Railways and the economic development of Western Europe, 1830–1914* (pp. 28–48). London: Macmillan.

Casanas Valles, M. (1977). El ferrocarril en Espana, 1844–1868. Consideraciones en torno a una crisis. *Investigaciones Económicas, 4*, 39–68.

Comın Comın, F., Martın Acena, P., Munoz Rubio, M., & Vidal Olivares, J. (1998). *150 años de historia de los ferrocarriles españoles. Volumen I.* Madrid: Anaya.

Comision de ingenieros. (1864). Ante-proyectos del Plan General de Ferro-carriles. Memoria. *Revista de Obras Públicas, 12*, 198–204 & 208–215.

Congreso de los Diputados. (1850). *Información parlamentaria hecha por la Comisión de Ferro-carriles nombrada por el Congreso de los Diputados en 10 de enero de 1850.* Madrid: Imprenta Nacional.

De Diego Garcıa, E. (1983). El ferrocarril: La Habana-Guines. *Cuadernos de Historia Moderna y Contemporánea, 4*, 59–77.

De Riquer, B. (1994). Aproximacion al nacionalismo espanol contemporaneo. *Studia Historica. Historia Contemporánea, 12*, 11–29.

De Rus, G., & Inglada, V. (1997). Cost-benefit analysis of the high-speed train in Spain. *Annals of Regional Science, 31*, 175–188.

De Rus, G., & Roman, C. (2006). Analisis economico de la lınea de alta velocidad Madrid-Barcelona. *Revista de Economía Aplicada, 14*, 35–80.

Derry, T.K., & Williams, T.I. (1960). *A short history of technology from the earliest times to AD 1900.* Oxford: Clarendon Press.

Docwra, G.E., & Kolsen, H.M. (1989). Transport and Australian federalism, 1901–1972. *Journal of Transport History*, *10*, 59–73.
Dunham, A.L. (1941). How the first French railways were planned. *Journal of Economic History*, *1*, 12–25.
Dunn, J., & Perl, A. (1994). Policy networks and industrial revitalization: High Speed Rail initiatives in France and Germany. *Journal of Public Policy*, *14*, 311–343.
Elliot, J.H. (1963). *Imperial Spain*. London: Edward Arnold.
Fayard, A., Gaeta, F., & Quinet, E. (2005). French motorways: Experience and assessment. *Research in Transportation Economics*, *15*, 93–105.
Fenoaltea, S. (1983). Italy. In P. O'Brien, *Railways and the economic development of Western Europe, 1830–1914* (pp. 49–121). London: Macmillan.
Fernandez de Mesa, T.M. (1755). *Tratado legal y político de caminos públicos y possadas*. València: Joseph Thomàs Lucas en la Plaza de las Comedias.
Fleming, F.J. (1991). Profits and visions: British capital and railway construction in Argentina, 1854–1886. In C.B. Davis, K.E. Wilburn, & R.E. Robinson (Eds.), *Railway Imperialism* (pp. 71–84). New York: Greenwood Press.
Flichy, P. (1995). *Dynamics of modern communication. The shaping and impact of new communication technologies*. London: Sage. (Originally published in French. Paris: Éditions La Découverte, 1991)
Fremlindg, R. (1983). Germany. In P. O'Brien, *Railways and the economic development of Western Europe, 1830–1914* (pp. 121–148). London: Macmillan.
Gómez Mendoza, A. (2001). Madrid, centro de la red de comunicaciones. *Arbor*, *169*, 343–358.
Gray, I. (2009). Maintaining the power of central governments: Regional land transport in the Australian Federation, 1850–2007. *Journal of Transport History*, *30*, 22–39.
Greco, A., & Ragazzi, G. (2005). History and regulations of Italian highways concessionaires. *Research in Transportation Economics*, *15*, 121–133.
Hawke, G., & Higgins, J. (1983). Britain. In P. O'Brien, *Railways and the economic development of Western Europe, 1830–1914* (pp. 170–202). London: Macmillan.
Heinisch, R. (1992). High Speed Trains in Germany. *Rail International*, *23*, 23–24.
Herranz Loncán, A. (2008). *Infraestructuras y crecimiento económico en España (1850–1935)*. Madrid: Fundación de los Ferrocarriles Españoles.
Lluch, E. (1996). *La Catalunya vençuda del segle XVIII. Foscors i clarors de la Il·lustració*. Barcelona: Edicions 62.
Madrazo, S. (1984). *El sistema de transportes en España 1750–1780*. Madrid: Ediciones Turner.
Mateo del Peral, D. (1978). Los orígenes de la política ferroviaria en España (1844–1877). In M. Artola, R. Cordero, D. Mateo, & F. Menéndez, *Los ferrocarriles en España 1844/1943. Tomo I. El Estado y los ferrocarriles* (pp. 31–159). Madrid: Servicio de Estudios del Banco de España.
Menéndez Pidal, G. (1951). *Los caminos en la Historia de España*. Madrid: Ediciones Cultura Hispánica.
Nadal, J. (1975). *El fracaso de la revolución industrial en España, 1814–1913*. Barcelona: Ariel.
Pascual Domènech, P. (1999). *Los caminos de la era industrial. La construcción y financiación de la Red Ferroviaria Catalana (1843–1898)*. Barcelona: Edicions de la Universitat de Barcelona.
Pérez Touriño, E. (1994). Efectos del Plan. Realizaciones y balance socioeconómico. *Estudios de Transporte y Comunicaciones*, *65*(separata), 9–27.
Ringrose, D.R. (1994). Madrid, capital imperial (1561–1833). In S. Juliá, D. Ringrose, & C. Segura, *Madrid. Historia de una capital* (pp. 153–314). Madrid: Alianza Editorial, 1994.
Ringrose, D.R. (1996). *España, 1700–1900: El mito del fracaso*. Madrid: Alianza Editorial.
Rothengatter, W. (2005). Motorways and motorway finance in Germany and Austria. *Research in Transportation Economics*, *15*, 75–91.
Smith, A. (1776). *The wealth of nations* (3rd ed.). London: W. Strahan & T. Cadell, 1784.
Tortella Casares, G. (1973). *Los orígenes del capitalismo en España: Banca, industria y ferrocarriles en el siglo XIX*. Madrid: Tecnos.
Uriol, J.I. (1977). Los viajes por la posta en el siglo XVIII y en los primeros años del siglo XIX. *Revista de Obras Públicas*, *124*, 837–856.

Uriol, J.I. (1978). Las carreteras y los canales de navegación en los reinados de Fernando VI y Carlos III. *Revista de Obras Públicas, 125*, 533–546.
Uriol, J.I. (1985). Las calzadas romanas y los caminos del siglo XVI. *Revista de Obras Públicas, 132*, 553–563.
Vicens Vives, J. (1952). *Aproximación a la Historia de España*. Barcelona: Vicens Vives.
Vickerman, R. (1997). High-speed rail in Europe: Experience and issues for future development. *Annals of Regional Science, 31*, 21–38.
Villuga, J. (1546). *Repertorio de caminos*. Medina del Campo.
Ward, B. (1762). *Proyecto económico, en el que se proponen varias providencias, dirigidas a promover los intereses de España, con los fondos y medios necesarios para su plantificación* (2nd ed.). Madrid: D. Joachín Ibarra, Impresor de Cámara de S.M.
Wotherspoon, G. (1979). The determinants of the pattern and pace of railway development in New South Wales, 1850–1914. *Australian Journal of Politics & History, 25*, 51–65.

States and start-ups: Public competitors in Australian communications

Jock Given

Institute for Social Research, Swinburne University of Technology, Melbourne, Australia

This article explores three state-owned or supported communications enterprises: the Pacific Cable, opened in 1902; Amalgamated Wireless Australasia (AWA), which launched international wireless telegraphy services in 1927; and AUSSAT, a domestic satellite system established in 1985. All were partnerships formed to build and operate new communications infrastructure in competition with an incumbent. Of each, the article asks: what problem needed to be solved? What institutional solution and technologies were chosen and why? How did incumbents respond? How did the enterprise perform and why did it end? It finds very mixed performance, both as investments of public money and as solutions to the problems they were set up to solve.

The return of the state

In April 2009, the Australian Government announced a new plan for a National Broadband Network costing AUS$43 billion (US$43 billion in mid-November 2010). A state-owned company is now building the 'fibre-to-the-premises' (FTTP) network to 93% of homes and workplaces. Announcing this future, the Prime Minister talked a lot about history. He called it 'the most ambitious, far reaching and long term nation-building infrastructure project ever undertaken by an Australian government ... Like the building of the Snowy Hydro, like the building of the Sydney Harbour Bridge, this is an historic act of nation building' (Rudd & Swan, 2009).

Around the world, some governments were already supporting investments in 'next generation' fibre access networks (Bauer, 2010). By late 2007, Huigen and Cave (2008, p. 714) say the question had become 'not so much whether a massive increase in bandwidth over fixed access networks is necessary, but rather when and how it will be realised'. Government intervention got a boost from the global economic and financial crises. Infrastructure spending, especially for broadband, was identified as one of three broad fiscal/structural reforms that could yield a 'double-dividend' by raising aggregate demand in the short-term and aggregate supply in the long-term (OECD, 2009a; Reynolds, 2009).

This was a significant shift from the liberalisation and privatisation consensus of the 1980s and 1990s, though not a complete turnaround. In many countries, privatisation stalled before it was complete, leaving big telecommunications companies like NTT, France Telecom, Deutsche Telecom and Telia Sonera with large public share-holdings (OECD, 2009b, pp. 43–46). State-ownership of communications enterprises had a long history. Writing before the telegraph and the telephone, Adam Smith (1778/1938, pp. 300–301) was generally unenthusiastic about states running businesses. He deplored princes becoming 'adventurers in the common branches of trade ... No two characters seem more inconsistent than those of trader and sovereign'. Yet even he argued the post office was 'properly a mercantile project', perhaps the only one 'successfully managed by, I believe, every sort of government'. Smith thought there was 'no mystery in the business. The returns are not only certain, but immediate'. In the early decades of the International Telecommunications Union, nationalisation or complete control of the telegraph system was 'always an unwritten prerequisite for membership'. This excluded the United States, though its companies accepted ITU regulations and participated in revising them, and, until telegraph nationalisation in 1870, Great Britain (Codding, 1972, pp. 38–42).

Private participation in communications, however, was common, especially in international communications. The Eastern Telegraph group dominated submarine cables in the late nineteenth and early twentieth centuries (Headrick, 1991). Early communications corporations often blended private investment and public goals. The former head of the United States Post Office's railway mail service, Theodore Vail, joined AT&T in 1907, where he used the language of 'universal service' – 'one system, one policy, universal service' – to help transform the telephone industry into a state-sanctioned but still private monopoly (John, 1999). The Radio Corporation of America (RCA) was a private company, but creating it in 1919 was a government-sanctioned act of industrial nationalism (Aitken, 1985, pp. 302–431). Britain's Foreign Office took effective control of the newsagency Reuters during the First World War (Read, 1999; Putnis, 2010). The BBC was first a private, then a public, monopoly in the new field of radio broadcasting in the 1920s.

At a time of renewed interest in industrial policy ('The global revival', 2010), this article explores three state-owned or supported communications enterprises: the Pacific Cable opened in 1902, Amalgamated Wireless Australasia (AWA), which launched international wireless telegraphy services in 1927, and AUSSAT, a satellite system established in 1985. The Pacific Cable and AWA provided new connections between Australians and the world; AUSSAT connected Australians to each other and eventually to some island neighbours. All were partnerships formed to build and operate new communications infrastructure in competition with an incumbent. Of each, the article will ask: what problem needed to be solved? What institutional solution and technologies were chosen and why? How did incumbents respond? How did the state enterprise perform – both in solving the initial problem and as a commercial investment of taxpayers' funds? Since none of these three enterprises still exists in anything like its original form, the paper will ask how and why they ended.

What problems needed to be solved?

The Pacific Cable was established to create competition in international telegraph messaging between Australasia and the Northern hemisphere. AWA's wireless services, launched 25 years later, were designed to increase it. Competition was expected to reduce prices and improve service quality and reliability. Infrastructure independent of existing networks would be invaluable during outages caused by technical failure, bad weather or war. AUSSAT, officially, was intended to provide new and better services to remote parts of Australia. The competition that satellites could have given to the incumbent's terrestrial networks was initially constrained by the institutional structure chosen and the regulation applied to it. All three projects served political ends – the Pacific Cable and AWA's services were to be empire-binding, AUSSAT nation-binding. By making international telegrams affordable for individuals as well as businesses, the price reductions promised by international telegram competition would foster political ends as surely as AUSSAT's later extension of facilities to unserved areas.

Complaints about the quality and cost of international telegrams to and from Australia and New Zealand began almost as soon as services commenced in the 1870s. The London-based Eastern Telegraph group extended its England-India submarine cable further east to Java and Port Darwin. There it connected to a landline south across Australia to Adelaide. An 'overland telegraph line' through the centre was chosen ahead of a landing on Cape York and a line down the populated east coast to Sydney and Melbourne because the South Australian colonial administration made an irresistible offer to pay the entire cost of the overland line. This would make Adelaide the continent's global electronic gateway. The individual heroism of its constructors turned the Overland Telegraph Line into an historic achievement, but the service was expensive and as unreliable as critics of the route had feared. Yet even when traffic outstripped capacity, a second line up the coast was rejected. The overland line was duplicated, then the submarine cable to Darwin. Customers remained dependent on Eastern's monopoly (Headrick & Griset, 2001; Horgan, 1986; Johnson, 1903; Putnis, 2008).

Genuine competition required a new line to London independent of Eastern's lines through the Indian Ocean and Mediterranean Sea. This brought opportunities as well as costs. A Pacific cable would need to traverse unprecedented underwater distances and depths. By linking the Australian colonies to New Zealand, Norfolk Island, Fiji and Canada, however, interconnecting to the new Canadian Pacific Railroad telegraph line and an Atlantic cable to London, it could be part of a global 'All-Red Line' that never touched non-British territories. This imperial motivation did not wholly favour a Pacific line, because competition also threatened Eastern. The great British communications company was itself a tool of empire, its global network of cables heavily used for secure defence communications and its fleet of repair ships standing ready to inflict damage on the oceanic infrastructure of hostile powers. But the All-Red goal suited Joseph Chamberlain's New Imperialism in the late 1890s and its influence was highlighted by the eventual choice of British-controlled Fanning Island (now Tabuaeran, part of Kiribati) rather than the more economically useful Hawaii for the first landing point west of Vancouver.

Wireless promised even tougher competition for the international telegram business. Marconi's Wireless Telegraph Company opened a trans-Atlantic commercial service in 1907, six years after first transmitting a Morse Code signal across the ocean (Baker, 1970). In Australia from 1911, the former Marconi ship's wireless operator Ernest Fisk imagined a wireless service connecting the south-east of Australia to Britain, two places almost as far away as was possible on the earth's surface. During the First World War signals were fortuitously received on the west coast from Telefunken's high power station in Berlin, but Sydney and Melbourne were where the customers were. Fisk became managing director of AWA in 1917, four years after it was created by merging the local businesses of Marconi and

Germany's Telefunken. Direct messages were first received in Sydney from Marconi's Carnarvon station in September 1918.

In the language of contemporary telecoms policy, a direct wireless service would be a disruptive, facilities-based competitor for the wireline duopoly, Eastern and, since 1902, the Pacific Cable. AWA had economics, defence, politics and emotion on its side. It promised to reduce the submarine cable companies' rates by a third. Expensive new transmission and reception stations would be required at each end, but cable operators had to maintain thousands of miles of undersea cables with vastly less capacity than the optic fibre used today. Long distance wireless communication was vulnerable to atmospheric disturbance but not to severing by hostile military powers. A direct link with London resonated strongly in a Dominion distant from the heart of the British Empire, especially so soon after a war that had claimed 60,000 of its people (Australian House of Representatives, 1921).

In the late 1970s, a national satellite system offered different things to different interests (Paltridge, 1989). The main political justification was 'greater equity in the distribution of communication services throughout Australia' (Barr, 1986). Like the Pacific Cable, it did not involve wholly new technology. Australia had a 2.5% stake in the Intelsat global satellite system and built its first earth station in 1966. Remote viewers were able to watch television for the first time in 1980 via large dishes and low-power terrestrial retransmitters. The appeal of a national satellite system like those in the US, Canada and Japan (Commonwealth Government Task Force, 1978) lay in being able to optimise its design for particular domestic services. Receivers could be smaller and cheaper than for Intelsat's spacecraft. In theory, users in remote mining towns and Indigenous communities could receive services with the same equipment, costing the same amount, as people in Sydney (Working Group, 1979).

Intense conflict arose about the forms of service the satellite could carry. The Minister said it was designed as a point-to-multipoint system 'for applications in which satellite is the only solution, or the most cost-effective solution, or the only ideal solution'. These included broadcasting, air navigation, and high speed data and voice networks in situations where the required data rate could not be supplied by the terrestrial incumbent quickly enough or at all (Staley, 1985). But by offering an alternate way of delivering services like TV programme distribution, data transmission and voice telephony, a domestic satellite became a 'Trojan Bird' for those wanting to end the country's telecommunications monopoly (Reinecke & Schultz quoted in Hazlehurst, 1990). Its most enthusiastic supporter was a commercial TV operator, Kerry Packer, who commissioned RCA's Donald Bond to write the report that initiated government interest in the project. He wanted an alternative to Telecom's union-dominated workforce for distributing programmes around the National Nine Network and the ability to broadcast direct to country TV viewers (Hazelhurst, 1990).

What institutional solutions and technologies were chosen and why?

Because competition was a major reason these three projects were promoted, they had to be undertaken by competing institutions. The Pacific Cable had to be built and operated by someone other than Eastern; the wireless services by someone other than the cable duopoly; the domestic satellite by someone other than Telecom Australia. The state got involved in different ways. The Pacific Cable was wholly owned by four different nation states. AWA became a partnership between public and private investors where the Australian Commonwealth held a bare majority of shares but a minority of the directors. AUSSAT was conceived by a conservative government to be like AWA, a roughly equal partnership between the state and private investors. Wholly state-owned at the start, the plan to sell down the public stake to 51% never eventuated once the government changed and the financial performance made a float impossible.

The Pacific Cable became a state-owned enterprise because it was hard to imagine a fully commercial enterprise risking head-to-head competition with the globally dominant and politically well-connected Eastern, especially for a cable terminating in Canada rather than the United States. While supporters like Canada's Sandford Fleming (Thompson, 1990) produced forecasts showing it would be profitable based on reasonable estimates of traffic growth and likely market share, they also stressed that only a public enterprise could afford to pursue the public goal of setting cheaper prices. Fleming wrote to Chamberlain in 1898:

> I am satisfied that the Pacific Cable would prove to be the entering wedge to remove forever all monopoly in ocean telegraphy, and free the public from excessive charges; that it would be the initial link in a chain of state cables encircling the globe, with branches ramifying wherever the British Empire exists (Johnson, 1903, p. 480).

Drawing together governments from four places – New Zealand, Canada, Britain and Australia – greatly complicated the arrangements, but emphasised the political dimension. The imperial mission became the organisational form, a London-based company managed by a board with representatives from the four governments, financed by a £2 million loan from the UK National Debt Commissioners repayable over 50 years. Crucially, all the early discussions in Australia were with the six separate colonial governments, and occurred alongside the negotiations that led to their federation in 1901. The financial structure eventually agreed, whereby profits and losses were shared by the partners after deducting the annual interest and capital repayment on the loan plus a £30,000 contribution to a Renewal Fund, reflected the pre-Federation disagreements about the whole project. The colonies of Queensland (which eventually hosted the initial landing point, at Southport), New South Wales, Victoria and New Zealand supported it and wore 2/18ths each, giving Australia 6/18ths after federation. Tasmania, South Australia and Western Australia opposed it and stayed away. The UK and Canada bore 5/18ths each.

A public/private partnership was chosen for the international wireless services in the 1920s rather than a wholly public or private entity for several reasons. Patents for the technology were controlled by the Marconi Company and its exclusive Australasian licensee, AWA. Rather than establish a new company, the Commonwealth chose to recapitalise AWA. It committed £500,001 to acquire a bare majority stake in a company capitalised at £1 million. Marconi remained the largest private shareholder. Prime Minister Billy Hughes was no fan of the Post Office, the most likely operator of a state-owned wireless service. A former waterfront union official expelled from the Labor Party over his war-time support for conscription, Hughes had migrated from the political left to the capitalist right by the early 1920s. Here, a commercially-managed partnership seemed more attractive than the nationalisation urged by Labor. The state could afford the capital injection and the business case forecast good dividends. For AWA, it brought the state inside the enterprise, a high-risk strategy to end the Post Office campaign to extend state control of wireless into areas where AWA saw future growth opportunities. The board comprised three government representatives, three representatives of the private shareholders and a seventh, nominated by the other six. Hughes became 'seventh director' himself, in practice delivering effective control of the company to its managing director once he was forced to resign as Prime Minister.

There were recent precedents for public/private partnerships. A bare majority stake in Anglo-Persian Oil (later Anglo-Iranian, still later BP) was acquired by the British Government in May 1914. British naval and industrial power had long relied on dependable supplies of coal, as they had on a global cable communications network. Oil, like wireless, offered some advantages to public and private users, but it required huge investment to convert power supplies or to build new oil-fired ships and factories. It also required new transport and storage facilities around the world. Securing access to cheap oil became an important goal for the First Lord of the Admiralty, Winston Churchill. He championed the investment in Anglo-Persian to maintain the independence of a cash-strapped company in a global market dominated by Standard Oil and Royal Dutch-Shell, and to secure part of the Royal Navy's oil supplies (Ferrier, 1982). The public/private Anglo-Persian entered into a further partnership with the Australian Government, the Commonwealth Oil Refineries. In the Australian Parliament, Prime Minister Hughes referred to this precedent but it was not seriously analysed.

The debate about the ownership of the domestic satellite occurred in the late 1970s and early 1980s as dissatisfaction with state-owned telecoms monopolies and duopolies intensified. The British Post Office's telecoms activities were transferred to a separate corporation in 1981, then privatised between 1984 and 1993, and Cable & Wireless subsidiary Mercury was granted a licence to compete with BT in Britain in 1982. A United States court finally settled the long-running anti-trust action brought by the Department of Justice against AT&T in a 'modified final judgment' that took effect in 1984, forcing divestiture of the company's local operating companies. In Australia, the Packer-commissioned Bond Report framed a domestic satellite as an opportunity to draw private investment into the sector. AUSSAT was established as a corporation wholly-owned by a government that expected it to pay taxes and dividends, and planned to sell down its interest to 51%.

This plan changed when a Labor administration was elected in 1983. A 25% stake was sold to Telecom Australia, the state retained 75% and board members who worked for commercial media companies were removed. This significantly altered the satellite corporation's competitive aspirations (Barr, 1986). The Liberal/National coalition had also commissioned a telecoms policy review that supported market liberalisation (Davidson, 1982). Although it rejected this review's major recommendations, it authorised private earth stations for the new satellite system, allowing Packer and other TV operators to free themselves from Telecom's infrastructure for programme distribution. While initially opposing wider liberalisation, the Labor Government removed some restrictions on the services that could be operated from the ground stations (Barr, 1986).

How did incumbents respond to the new enterprises?

Incumbents mounted similar campaigns to resist their new competitors. While the projects were being planned, they lobbied fiercely against them, arguing the new capacity was not necessary, the technology wouldn't work effectively and the services would not be profitable. Once it was inevitable, however, they sometimes offered to build the new infrastructure themselves, and acted to shore up and expand their businesses before it was built. Once it was operating, they worked to destroy it, hoping to ensure it never happened again, or at least not with government backing.

Eastern argued the relevant part of the Pacific was too deep for a submarine cable, the coral reefs where the big lengths of cable would have to terminate were too rough, and the route was too far from shipping lanes and coaling stations (Horgan, 1986). When Marconi himself asked a cabinet committee in April 1924 for six months to further develop and demonstrate the new directional short-wave 'Beam' technology that was delivering him such remarkable experimental results, and which he thought should replace the long-wave technology already endorsed for international telegraph services, the specialist technical adviser, W.J. Eccles, suggested six years (Eccles, 1924). In the 1970s, Telecom Australia told the Minister that low powered satellites could not yet deliver television pictures that would be viewable in homes via small, low cost receivers (AUS$1000 became the target) (Staley, 1985).

Eastern had skilfully extracted concessions, subsidies, traffic guarantees and exclusive rights for decades from governments around the world. Australasian governments subsidised the Overland Telegraph Line, the duplication of the Java/Darwin cable that connected to it (1879) and the cables between Tasmania and the mainland (1869) and Sydney and New Zealand (1876). Each new arrangement presumed the service would otherwise be uneconomic, but the combined effect was to turn Eastern's prosperity into a public policy goal. Anything that threatened it could not be in the public interest. Sandford Fleming lamented at the Colonial Conference in 1886:

> Is it for a moment to be thought that Canada and Australasia are never to hold direct telegraphic inter-course because a commercial company stands in the way? ... Is the peace of the world to be endangered at the bidding of a joint stock company? (Quoted in Horgan, 1986, pp. 45–49)

The Pacific Cable Board complained about the likely impact of a wireless service on its ability to upgrade its facilities and pay a capital return to its government stakeholders. Telecom Australia argued that a satellite rival would cherry-pick its customers and undermine its ability to make services universally available.

But once the new services were inevitable, the threatened incumbents all invested heavily to meet the looming challenges. Eastern cut its rates in 1890 and then in 1897 offered to build a new cable to Australia's west coast from South Africa costing £2 million, provided the United Kingdom and Australian colonies paid annual subsidies totalling nearly £60,000. Just three weeks after all governments finally agreed to proceed with the Pacific Cable in July 1899, Eastern dramatically revised this offer. It would build the new line without subsidy and cut its rates on 1 January 1901 and again a year later, or immediately, if permitted to establish its own offices in the major cities to collect and distribute international telegrams (rather than using Post Offices) and to lease exclusive landlines to connect them. This split the soon-to-be-federated colonies and Pacific partners. Western Australia, South Australia and Tasmania signed up in April 1900; Victoria and NSW were torn; Queensland, New Zealand and Canada were opposed. NSW eventually signed as well, taking advantage of the rate cuts and allowing Eastern to establish a retail presence in

Sydney. Eastern's cable-laying subsidiary Telcon won the contract to build the Pacific Cable, signed the day before Federation (Horgan, 1986). Eastern's new Indian Ocean cable opened on 1 November 1901, more than a year before the Pacific Cable (Harcourt, 1987, p. 171).

Eastern and the Pacific Cable maintained their standard rates for more than two decades, although the Pacific Cable introduced a cheaper weekend lettergram rate in 1913 that was very popular especially during the war. Standard cable rates were lowered in 1924 when AWA's wireless service loomed and again two months before the service began in April 1927. The Pacific Cable responded to the wireless threat and to its own message delays by agreeing to the huge investment required to duplicate the cable in two stages. The cable south of Fiji was duplicated in 1923, and the much longer and more expensive northern section in 1926, at a total cost of around £2.75 million. There was heavy criticism of the Board's sudden decision to go ahead with full duplication just as a more cost-efficient wireless technology was about to be launched, especially by Canadians (PCB, 1922–26; PMG's Department, 1927). Anticipating extra traffic from its duplication, the Pacific Cable took an exclusive lease for a second landline across Canada from Vancouver to Montreal that was handed over just two months after the Australia-England wireless service opened and a year before the Australia-Canada service (PCB, 1919–28).

Telecom argued a domestic satellite system was too expensive a way to improve 'even the most pressing and difficult of remote area telecommunications services'. Instead, it developed a terrestrial telephone system for remote areas. By 1990, it had spent AUS$530 million (US$415 million at the time) on this, about the same amount AUSSAT had spent on its three satellites. One consultant said there was 'a considerable degree of scepticism in the satellite industry about Telecom's claims', and thought it 'fair to say that Telecom's attempts to substantiate these assertions have been neither clear nor completely convincing' (Aked, 1990, p. 56). Like other telcos around the world, it began replacing coaxial cable and microwave transmitters in its trunk routes with optical fibre in the early 1980s (Moyal, 1984, p. 365). Though expensive to lay, the huge capacity of fibre made it a much cheaper way to deliver some of the services where AUSSAT had hoped to be competitive.

How did the state-supported enterprises perform?

The Pacific Cable did not make profits until the First World War, then struggled with surging demand during it and the need for reinvestment after it to meet customer needs and the looming threat of wireless. Those wireless services were so immediately and dramatically successful in the late 1920s that they motivated a merger of cable and wireless interests in Britain, into which the Pacific Cable was sold. AUSSAT was a financial debacle, building up so much debt and demanding so much new capital for its second generation of satellites that it triggered a wholesale restructuring of the Australian telecommunications industry, including the privatisation of AUSSAT itself.

Those who had argued the Pacific was too deep for a submarine cable were wrong, both about just how deep it was and about how the cable would fare. Laying it between Fanning Island and Vancouver, the depth was never more than 3400 fathoms, not 12,000 as some had predicted in 1887. These still deep sections, however, proved particularly robust. Much more trouble was caused by the rough seas and rocks around the landing at Norfolk Island. After nearly two decades of operation, the only 'natural or accidental interruption' on the two longest sections between Fiji and Canada had occurred when the German cruiser Nürnberg landed on Fanning Island and cut the cables soon after the outbreak of the First World War (Harcourt, 1987).

Sandford Fleming's estimates of total Australasian international cable traffic for the early years proved accurate, but not the Pacific Cable's share of it. In 1903, Eastern had 83% of Victoria's international traffic and 74% of NSW's, clearly demonstrating the benefits of incumbency and the retail offices the former monopolist set up to attract customers. Pacific's market share grew after the federal government forced the closure of Eastern's Melbourne office in 1906, angered by its refusal to accept an expiry date for the pre-Federation agreements (PMG's Department, 1911). On the eve of the war, Eastern still had considerably more than half the Australasian international traffic (excluding intermediate traffic). When it peaked in 1917/18 at over 16 million words, nearly three times the level in 1913, Eastern had two-thirds of it (PMG's Department Annual Reports, 1911–47).

The Pacific Cable introduced a cheap 'Weekend Lettergram' rate in 1913 that was hugely popular especially for short messages to serving soldiers, sailors and nurses. It accounted for more than 40% of Pacific's traffic in 1915/16 but cannibalised the full-rate business. Private codes, used before the war to condense messages, were banned in wartime, but businesses got better value out of the new, cheap plain language services. Eastern never matched the Weekend Lettergram rate, fearing it would not be able to cope with demand. The fear proved well-founded. Pacific eventually had to suspend the service to ensure priority for full-rate and government messages and only restarted it in 1921 (PCB, 1915–23).

Supporters of the Pacific project noted that Australia's one-third share of the losses sustained in the early years (just under £30,000 in the first full year, 1903/04) was less than the annual subsidy it had paid to Eastern in the 1880s and 1890s to support duplication of the Java-Darwin cable (£32,400) (Livingston, 1996). For this, the Australian public was getting a stake in an asset, which it didn't by subsidising Eastern's cable. The Pacific's losses declined steadily and it first made a profit in 1915/16. Total accumulated losses by this stage had reached nearly £623,000. Annual surpluses earned from then were not distributed to the partners, but used to pay off the original loan ahead of schedule and to build up the Renewal Fund, first to replace the sharp war-time decline in the value of its investments in stocks and bonds and then to provide for the anticipated cost of duplication.

Opening in April 1927, about six months after the first direct wireless service between England and Canada, the England-Australia wireless telegraph service quickly took market share from the Eastern and Pacific cables. The higher frequency short-wave directional 'Beam' technology required less power than long-wave, reducing capital and operating costs. Its average capacity of 250,000–300,000 telegraphic words per day was around twenty times the capacity agreed in 1922. The Australia-England rates were lower than those agreed five years earlier – 1s. 8d. a word for full rate messages, 4d. below the promised 2s. rate which was itself a one-third reduction on the 1922 cable rate. Much lower rates were set for press cables. AWA opened new public Accepting Offices in Melbourne and Sydney. Across the rest of the country, any Post Office could accept a telegram to England and connecting destinations marked 'via Beam' (Beam Wireless to England, c. 1927). Further imperial wireless services connecting England to South Africa and India opened later in 1927, and AWA opened a service to Montreal on 16 June 1928 (Vyvyan, 1933).

In 1928/29, the first full year of operation of both the Australia-England and Australia-Canada services, the wireless services took a remarkable 49% of Australia's international telegraph traffic, leaving the other half to be shared by the old rivals, Eastern and the Pacific Cable (PMG's Department Annual Reports). Having paid no dividend since 1922/23 while it was building the infrastructure, AWA declared one of 8% on the paid-up capital in 1929, a week after Wall Street's Black Tuesday. Total traffic slumped 25% over the next three years and the dividend fell to 6%, but it returned to 8% in 1932/33 and rose steadily to 12% in 1936/37 and 1937/38, above the 10% promised when the plan was first presented to the Parliament in 1921 (AWA Annual Reports). The market value of AWA's shares grew 3.1-fold from the time of the recapitalisation in 1922, to January 1945. This was better than Australian stock prices generally, which grew 2.8 times over the period. It is impossible to tell from the available data whether this performance was generated by the international wireless services that motivated the original Commonwealth investment, or the new areas of the wireless business that opened up after it, particularly radio broadcasting and manufacturing (Given, 2007).

AUSSAT's first expenses were incurred in 1981/82 and revenues arrived in 1985/86. Its AUS$100 million (US$70 million at the time) capital was fully called by 1985 when the first two satellites were launched. The system eventually comprised three geo-stationary satellites with footprints covering Australia, New Zealand, Papua New Guinea and the south-west Pacific, eight major earth stations – one in each of Australia's state capital cities plus Wellington and Auckland in New Zealand – and 15,000 smaller earth stations maintained by mining and exploration companies, hotels and clubs, news agency clients and remote communities. The main services offered were direct to home television broadcasts; distribution, relay and assembly of TV and radio programmes especially between the cities and country centres; a national aviation communications network of 200 earth stations at airports; and services for the Department of Defence (Joint Committee of Public Accounts [JCPA], 1994).

A pre-tax operating profit was earned for the first time in 1988/89, but by the end of 1990/91, accumulated losses totalled AUS$136m. The Government provided an extra AUS$100 million (US$79 million at the time) capital injection in mid-1990 but by then it was clear that the second generation of satellites would require further borrowings. There was 'little prospect of trading profitably under prevailing regulatory conditions' (JCPA, 1994). The financial problems were blamed on high start-up costs in a capital intensive industry; high financing costs, which took two-thirds of operating revenue in 1990/91 as interest rates soared; disappointing revenue once the strong growth to 1988/89 slowed; and restrictions on the commercial services the company could offer. A dissenting report by the Department of Finance representative on the government committee that first examined the satellite proposal had anticipated the financial difficulties and argued that cheaper options were available for delivering most of the services proposed (Commonwealth Government Task Force, 1978, pp. 132–136).

As performance sagged, many criticised the rules that prevented AUSSAT competing with the terrestrial incumbent Telecom Australia in services such as switched telephony. Leonard (1990, p. 36) said 'It was perhaps inevitable that the country of the emu and the platypus should design a bird with curiously clipped wings and then complain that it could not fly for profit'. Aked (1990, p. 57) however argued that although the commercial failure was hastened by Telecom's aggressive targeting of the limited telecoms services that AUSSAT could offer, allowing it to offer a wider range probably would not have saved what were essentially television satellites from commercial failure. They were simply not designed to be able to offer competing telecoms services at sufficiently attractive prices. The main benefit was coverage of the entire continent. This meant that once the satellites were operating, services could be made available everywhere, much more quickly than via terrestrial infrastructure.

How did the enterprises end?

Each of the three enterprises ended its life by being sold into larger enterprises as part of wider changes to telecommunications institutions and regulation. The wholly public Pacific Cable and AUSSAT were privatised; the public/private AWA wireless services were nationalised. The ends came progressively faster – 27 years after services commenced for the Pacific Cable, 19 for AWA, 7 for AUSSAT. Cable and Wireless was on the scene each time. It acquired the Pacific Cable, had its Australian cable assets nationalised after the Second World War alongside AWA's international wireless business and was a major shareholder in the company that acquired AUSSAT in the early 1990s.

The Pacific Cable ended as a discrete enterprise when the partner governments sold it to the newly-created colossus Cable and Wireless (initially Imperial and International Communications Ltd). This privatisation was part of a complete restructuring of Britain's international telecommunications services agreed in 1928. Eastern Telegraph had warned throughout the 1880s and 1890s that a Pacific cable would not be viable. It wasn't, though partly because of Eastern's own actions and the ambitious Pacific duplication project in the 1920s. Desperate for a solution once wireless savaged the cable business, the exhausted public partners turned to the old private incumbent to buy out the Pacific Cable for a fraction of the amount they had recently spent duplicating it. Resisting, fighting and finally outlasting the competition, Eastern Telegraph ended up owning it.

Within months of the Australia-England wireless service opening, Eastern approached Marconi to discuss a merger and the British Government convened a conference to discuss the state of international cable and wireless services. Eastern was commercially vulnerable but still politically well-connected. Marconi's short-wave wireless technology was triumphant, but his company was not. It did not operate the services in Britain, as its affiliate AWA did in Australia. Instead, it built the stations for the British Post Office and licensed its patents under a tough contract on which Marconi's later said it made a 'substantial loss' ('Marconi losses', 1927). The company also had wider financial problems resulting from what *The Times* described as 'reckless mismanagement', in a leader just weeks before the Australia-England wireless service opened ('Marconi losses', 1927). In July 1928, the conference recommended a merger along the lines already announced by Eastern and Marconi (Barty-King, 1979; Imperial Wireless and Cable Conference, 1928; Harcourt, 1987). It also recommended leasing to the new entity the British Post Office wireless stations (including the British end of the AWA Australia-England service), and transferring to it the state-owned Pacific Cable and the Post Office Atlantic cables.

The Pacific Cable was sold to the new entity in 1928/29 for about half the cost of the recent duplication. By March 1922, the Renewal Fund had risen to almost £1.8 million, and most of the healthy profits each year were directed to it. Over the next four years, £2.75 million was spent on duplication, but the Renewal Fund still stood at £357,000 in March 1928, two years after the project was completed. Allowing for the £1.2 million of the original £2 million debt still outstanding when the Pacific Cable was sold, the £516,000 sale price (calculated on 1 April 1928) valued the assets, including the cable stations and a Sydney CBD property as well as the cables, at just under £1.4 million. The partner governments split the sale proceeds in the same shares they had borne the losses accumulated in the early years and never repaid (PCB, 1929–42). The Australian Government initially accepted but eventually rejected a similar fusion of international cable and wireless interests within AWA. AWA and Cable and Wireless (taking over Marconi's shareholding in AWA) arranged a 'joint purse' agreement under which traffic could be shifted between the two networks, revenue was shared, and the parties agreed not to 'adopt any methods by way of canvassing or otherwise calculated to withdraw traffic from one party for the benefit of the other' (Imperial Communications Advisory Committee [ICAC], 1937).

Public/private ownership of AWA's international wireless communications only ended after the governments of Britain, its Dominions and India agreed to re-organise the imperial communications system after the Second World War by nationalising their various private and part-private companies. These included AWA in Australia and Cable and Wireless in Britain. It took a co-ordinated campaign in several countries with Labor governments and a cheque for £1.4 million from the Australian Government to buy the relevant assets from AWA (the whole amount went to the private shareholders: Martin, 2010). This was returned to shareholders as a capital repayment, of which half went to the Government as the holder of half the shares (Harcourt, 1987). AWA became purely a manufacturing company but the Commonwealth retained its stake until sold by the Menzies' Liberal/Country Party coalition government in 1951. International telecommunications remained state-owned until merged with the domestic incumbent then privatised progressively from the mid-1990s.

Throughout the 1930s, AWA resisted every effort at nationalisation. The Labor Party never altered its view that international wireless should be a public monopoly, like domestic telecommunications. Soon after the international services opened, the conservative side of politics came to this view as well and it was bipartisan policy during the 1930s, though not implemented. AWA's managing director was close to the Attorney-General in the Lyons administrations, Robert Menzies, and the company insisted massive compensation would have to be paid if its contract with the Commonwealth was unlawfully terminated, so as to deprive it of valuable rights. Menzies' Cabinet rejected a plan for the Post Office to take over the AWA services in 1940, agreeing only to 'the principle of co-ordination of communication services during the period of war' (PMG's Department, 1942). Although occurring under a Labor Government, the post-war nationalisation was supported in the Parliament by all sides.

AUSSAT's mounting debt precipitated a crisis in telecoms policy. No sooner had the government overhauled the regulation of the industry in 1989, allowing competition in 'value-added services' though not in the basic network, than it was forced to confront the parlous financial state of its satellite company. It decided to introduce network competition and private capital – steps it rejected just two years earlier – and to restructure the major telecoms organisations. The state-owned domestic and overseas telecoms corporations Telecom Australia and OTC would be merged but remain state-owned; AUSSAT would be sold to a private company licensed as the country's second carrier. Optus Communications, owned by Cable and Wireless, Bell South, a domestic transport company and local financial institutions, was selected ahead of a consortium of American 'Baby Bells', Hong Kong's Hutchison

Whampoa, and France Telecom. The AUS$800 million (US$615 million at the time) fee paid by Optus for the new licence roughly matched the AUS$799 million in debt accumulated by AUSSAT and paid off by the government before the sale, although deferred instalments reduced its net present value at the time of sale to AUS$700 million. Optus became Australia's new telecommunications carrier in January 1992. Cable and Wireless eventually took control of the company but then sold out to the current owner, Singapore's state-owned incumbent, Singtel. Optus's satellite business now performs strongly, operating five spacecraft with footprints covering Australia, New Zealand, Asia, Hawaii, Norfolk Island, Papua New Guinea, Lord Howe Island, Cocos Island, Christmas Island and McMurdo Sound.

Conclusions

These three enterprises are examples of the public sector financing competitive communications infrastructure and enterprises. As investments, it is tempting to interpret them as 'two funerals and a wedding'. Public investors suffered big losses on the Pacific Cable and AUSSAT but did well out of AWA. On closer examination, the picture is more complicated. The Pacific Cable was hit hard at the start by an aggressive incumbent that also had state support, and, at the end, by a disruptive technology. In between, it provided a good service and eventually earned steady profits. AUSSAT too faced an aggressive competitor and was partly a victim of policy contradictions that were not resolved until it was too late for the satellite company. The public/private AWA's share of the market for international wireless telegraphy makes it seem like a marriage made in heaven. In practice, the mixed enterprise lacked political support for most of its life and it is impossible to assess the financial performance of its different businesses.

Assessing performance is further complicated by disagreements about the rationale for establishing and maintaining the three enterprises. Paltridge's (1989, p. 8) observation about AUSSAT's first generation of satellites, that 'The measure of [its] success or failure ... will depend on who answers the question', could equally be applied to the Pacific Cable and AWA's wireless services. All were intended to deliver competition; their proponents promised they would do it profitably. AWA delivered both competition and profits. The Pacific Cable delivered price and facilities competition but ultimately at a heavy cost to shareholder returns. AUSSAT delivered losses and not much competition, but its failure was the trigger for more competition than ever before, precisely the outcome some of its critics had wanted to resist.

The promises of profitability were always tendentious. Purely commercial investors were unlikely to be attracted to ambitious projects opposed by powerful incumbents unless they got subsidies or regulatory guarantees. The state became involved because it appeared to be the only way to make the projects happen. Governments said they would make money, but their underlying hope may have been that they would not lose too much of it, while achieving other goals like lower prices, alternate infrastructure and more widely available services.

Both the founding and the dissolution of each enterprise were influenced by prevailing political ideologies and circumstances. The Pacific Cable was a creation of New Imperialism near the zenith of empire; its sale to Cable and Wireless was a consolidation partly to arrest imperial decline. Direct wireless services were an instrument of increasingly assertive Dominions; AWA's public/private structure and ultimate nationalisation were partly hangovers from war-time modes of operation.

AUSSAT spent its life on the cusp of two policy paradigms. Its establishment was a victory for emerging liberalisation; its ownership and restricted commercial charter were wins for the vanishing paradigm of state-ownership and monopoly. But although sometimes cloaked in ideology and principle, state-ownership or part-ownership of these three enterprises was essentially pragmatic, a way to get things done. Over time, governments that had not been part of the original deals found themselves uncomfortable with at least some of their implications.

One of the most important implications was the ongoing need for capital. For the Pacific Cable and AUSSAT, big new capital demands were decisive in finally unravelling political support. AWA found it had more capital than it needed but it also had other places to deploy it. Governments didn't want to be half-owners of things like manufacturing facilities and commercial radio stations but found themselves unable to unravel the structures. They lacked the ultimate control they imagined their bare majority shareholding gave them, or even the ability to wind AWA up, until other governments across the British world shared their resolve. In all three cases, state-ownership of existing institutions gave governments a big say in the shape of the institutions and industry that replaced them.

State-ownership tended to constrain commercial opportunities. Each of the three remained essentially a single-technology enterprise – cable, terrestrial wireless, satellite. This was much more of a handicap for the Pacific Cable and AUSSAT than it was for AWA, well-positioned in the middle of the wireless boom of the 1920s and 1930s. Indeed, Fleming, Merrett, and Ville cite AWA as an example of a 'prime mover' business that 'established an early dominance and maintained that position throughout the twentieth century'. The company was well-placed to offer the highly integrated production and services that evolved in the wireless business because of its domination of the early and evolving wireless technologies (Fleming et al., 2004, pp. 74 and 92–93). It was able to sustain this technical dominance largely by internal expansion, rather than more politically-risky acquisitions. Tellingly, each of the three single-technology state-supported enterprises was acquired by an organisation that encompassed both cable and wireless technologies. Single-technology enterprises helped to force change, dramatically so in the case of the wireless telegraph services, but further technological transformations (wireless, submarine cable telephony, optical fibre) eventually undermined their advantage.

Finally, each project shows the influence of individuals – those who championed them, including Sandford Fleming, Ernest Fisk, Billy Hughes and Kerry Packer, as well as those who resisted them, like Eastern Telegraph's John Denison Pender, whose sudden death in 1899 removed the Pacific Cable's most articulate and powerful opponent just as it was being considered by a UK Parliamentary Committee (Horgan, 1986, pp. 156–157). Led by Ernest Fisk from 1917–44 AWA is another example of the many leading Australian firms in the first half of the twentieth century that Fleming et al. (2004, pp. 72–73) argue were 'dominated or controlled by a particular entrepreneur or small group of decision-makers ... before the "managerial" firm belatedly began to supersede the "entrepreneurial" model ... after 1950'.

Promoting their visions, the proponents of the Pacific Cable, direct international wireless services and a domestic satellite all spoke about history, citing the transformative impacts of older communications projects to inspire support for contemporary ones. Announcing the final stage of the national satellite system in 1982, the minister spoke of the 'vision, courage and resourcefulness' of those who built the Overland Telegraph Line 110 years before. 'Undoubtedly this splendid enterprise will have as profound an effect on Australian life ... as did that single strand of galvanised iron wire swinging from 40,000 poles between Darwin and Adelaide in 1872' (Sinclair, 1982). Supporters of the National Broadband Network, similarly, place this twenty-first-century infrastructure alongside the Overland Telegraph Line of the nineteenth and the copper customer access network of the twentieth century as 'great public investments that have paid for themselves many times over in social, economic and productivity benefits' (Quigley, 2010). The case studies in this article show the mixed record of other public investments in Australian communications infrastructure, and the complexity of the factors that make neat conclusions about their impact so difficult to draw.

Acknowledgements

Research for this article was supported by the Australian Research Council, 2010–12 Discovery Project 1095061, 'Imperial Designs: Remaking the Institutions of Global Communications'.

References

Aitken, H. (1985). *The continuous wave: Technology and American radio, 1900–32*. Princeton, NJ: Princeton University Press.

Aked, A. (1990). AUSSAT as a telecommunications satellite. *Media Information Australia, 58*, 55–59.

Australian House of Representatives. (1921, October 5 and December 7). *Debates*.

AWA. (1914–46). Annual reports of directors. Sydney: AWA. (MSS 6275 Box 26) Mitchell Library, Sydney.

Baker, W.J. (1970). *A history of the Marconi Company*. London: Methuen & Co.

Barr, T. (1986). *The electronic estate: New communications media and Australia*. Ringwood, Australia: Penguin.

Barty-King, H. (1979). *Girdle round the earth*. London: Heinemann.

Bauer, J.M. (2010). Regulation, public policy and investment in communications infrastructure. *Telecommunications Policy*, *34*, 65–79.

Beam Wireless to England. Opening Date and Reduced Rates. (c. 1927). Beam Wireless – Tariff and Terminal Rates. (Series A461 Control C422/1/3) National Archives of Australia (NAA), Canberra.

Codding, G.A., Jr. (1972). *The International Telecommunications Union: An experiment in international cooperation*. New York: Arno Press. (Original work published 1952)

Commonwealth Government Task Force. (1978, July). *National communications satellite system report*. Canberra: AGPS.

Conroy, Senator S. (2009, April 7). *Joint media release with Prime Minister, Treasurer and Minister for Finance*. New National Broadband Network. Retrieved from http://www.minister.dbcde.gov.au/media/media_releases/2009/022

Davidson, J.A. (1982). *Report of the Committee of Inquiry into telecommunications services in Australia* (the 'Davidson Report'). Canberra: AGPS.

Eccles, W.J. (1924, April 24). Shorthand notes of a meeting of the Cabinet Committee on Report of Imperial Wireless Service, Whitehall Gardens. (HIS 119). Marconi Archive, Bodleian Library, Oxford.

Ferrier, R.W. (1982). *The history of the British Petroleum Company, vol. 1. The developing years 1901–32*. Cambridge: Cambridge University Press.

Fleming, G., Merrett, D., & Ville, S. (2004). *The big end of town: Big business and corporate leadership in twentieth-century Australia.* Cambridge: Cambridge University Press.

Given, D.J. (2007). *Transit of empires: Ernest Fisk and the world wide wireless* (Unpublished doctoral dissertation). University of Melbourne, Melbourne, Australia.

The global revival of industrial policy. (2010, August 7). *The Economist*, pp. 46–48.

Harcourt, E. (1987). *Taming the tyrant: The first hundred years of Australia's international communication services.* North Sydney: Allen & Unwin.

Hazlehurst, C. (1990). The dawn of the satellite era in Australia. *Media International Australia, 58*, 9–22.

Headrick, D.R. (1991). *The invisible weapon: Telecommunications and international politics 1851–1945.* New York: Oxford University Press.

Headrick, D.R., & Griset, P. (2001). Submarine telegraph cables: Business and politics, 1838–1939. *Business History Review, 75*, 543–578.

Horgan, J. (1986). *The Pacific Cable: A study of the 'connectional history' of Australia and Canada within the British Empire, 1872–1902* (Unpublished MA thesis). Sydney University, Sydney, Australia.

Huigen, J., & Cave, M. (2008). Regulation and the promotion of investment in next generation networks – a European dilemma. *Telecommunications Policy, 32*, 713–721.

Imperial Communications Advisory Committee (ICAC). (1937). Communications in Australia: Arrangements between I&IC and AWA. (POST 33/3660 f. 12). BT Archives, London.

Imperial Wireless and Cable Conference. (1928). *Report of the Imperial Wireless and Cable Conference. Cmd. 3163.* London: HMSO.

John, R.J. (1999). Theodore N. Vail and the civic origin of universal service. *Business and Economic History, 28*, 71–81.

Johnson, G. (Ed.). (1903). *The all red line: The annals and aims of the Pacific Cable Project.* Ottawa: James Hope & Sons.

Joint Committee of Public Accounts (JCPA). (1994, September). *The sale of Aussat, report 333.* Canberra: AGPS.

Leonard, P. (1990). Footprints down a narrow path: The regulation of AUSSAT. *Media International Australia 58*, 36–41.

Livingston, K.T. (1996). *The wired nation continent: The communication revolution and federating Australia.* Melbourne: Oxford University Press.

Marconi losses. (1927, March 16). *The Times*, p. 17.

Marconi's Wireless Telegraph Company. (1927, December 23). *The Times*, p. 16.

Martin, I. (2010, November 15). *Compensation paid to AWA for nationalisation of its international communications services 1946–47.* Communications Policy Research Forum, Sydney.

Moyal, A. (1984). *Clear across Australia: A history of telecommunications.* Melbourne: Thomas Nelson.

National Communications Satellite System Working Group. (1979). *National communications satellite system.* Canberra: AGPS.

OECD. (2009a). *Economic policy reforms: Going for growth.* Paris: OECD.

OECD. (2009b). *Communications outlook 2009.* Paris: OECD.

Pacific Cable Board (PCB). (1915–23). Finance. Paper file. (Series A2910 Control 430/1/72 PART 1) NAA, Canberra.

Paltridge, S. (1989, December). *Australian satellites: Promise, performance and the next generation.* Melbourne: Centre for International Research on Communication and Information Technologies.

PCB. (1919–28). Reports and Accounts. Paper file. (Series A458 Control F220/1) NAA, Canberra.

PCB. (1922–26). Duplication of Pacific Cables. Paper file. (Series A457 Control B207/1) NAA, Canberra.

PCB. (1929–42). Finance. Paper file. (Series A2910 Control 430/1/72 PART 4) NAA, Canberra.

PMG's Department. (1911–1947). *Annual reports.* Melbourne: Government Printer for the State of Victoria; Canberra: Government Printer.

PMG's Department. (1942, October 23). External cable and wireless services – proposed formation of new telecommunications company in Australia. (Series A571 Control 1944/1691) NAA, Canberra.
Putnis, P. (2008). The early years of international telegraphy in Australia: A critical assessment. *Media International Australia, 129*, 140–148.
Putnis, P. (2010). Reuters and the British Government – revisited. *Media History, 16*. 295–299.
Quigley, M. (2010, August 18). Charles Todd Memorial Oration, Sydney.
Read, D. (1999). *The power of news: The history of Reuters* (2nd ed.). Oxford: Oxford University Press.
Reynolds, T. (2009, May 19). *The role of communications infrastructure investment in economic recovery*. DSTI/ICCP/CISP(2009)1/FINAL. Paris: OECD.
Rudd, K., & Swan, W. (2009, April 7). Transcript of media conference at Parliament House, Canberra.
Sinclair, I. (1982, May 6). Ministerial Statement. House of Representatives Debates.
Smith, A. (1938). *The wealth of nations. Vol 2*. London: J.M. Dent & Sons/Everyman's Library. (Original work published 1778)
Staley, A.A. (1985). Reflections on the genesis of AUSSAT. *Media Information Australia, 38*, 9–12.
Thompson, G. (1990). Sandford Fleming and the Pacific Cable: The institutional politics of nineteenth century imperial telecommunications. *Canadian Journal of Communication, 15*, 64–75.
Vyvyan, R.N. (1933). *Wireless over thirty years*. London: George Routledge & Sons.
Working Group on National Communications Satellite System. (1979). *Working group report*. Canberra: AGPS.

How politics, economics, and institutions shaped electric utility regulation in the United States: 1879–2009

William J. Hausman[a] and John L. Neufeld[b]

[a]*Department of Economics, College of William & Mary, Williamsburg, VA, USA;* [b]*Department of Economics, University of North Carolina at Greensboro, Greensboro, NC, USA*

The history of electric utility regulation at both the state and national level from the beginning of the industry through the aftermath of the California energy crisis of 2000–01 is presented. That history was partly determined by the economics of the industry – on the supply side by its cost structure, network characteristics, and lack of storability – on the demand side by its price inelasticity for all but the largest consumers, and partly by politics. These factors influenced the institutions that were created to regulate the industry, a process also complicated greatly by US federalism. The intensity of regulation waxed and waned in response to real or perceived problems in the industry.

State commission regulation of utility rates

Under the form of regulation that has dominated the US electric utility industry for most of the twentieth century, a state regulatory agency both protects the monopoly status of private utilities and oversees the rates and rate structures that a utility can impose upon its users. Wholesale sales between utilities came to be regulated by a federal agency (Federal Power Commission) that uses a methodology similar to that used by state regulators. The industry, from the beginning, also included municipally-owned utilities and eventually state- and federal government-owned utilities that typically are not subject to outside regulation.[1] The complexity and decentralised control that characterises this system makes broad reform of the industry politically difficult to accomplish. Nevertheless, dramatic changes in utility regulation have occurred at times during the industry's history.

The methodology traditionally employed by state and federal regulatory commissions was largely determined by the first (1898) Supreme Court review of rate regulation by state commissions, Smith v. Ames (169 US 466), which concerned a Nebraska commission that had the power to reduce railway rates. In that decision, the Court concluded that businesses subject to rate regulation had a constitutionally-protected (based on the due process and equal protection clauses of the Fourteenth Amendment) right to receive a 'fair' return on the 'fair value' of its investment. The

Court furthermore ruled that this applied not just to the overall business of the utility but to each segment, such as passenger service and freight service. Under this rule, a regulatory commission had to determine: 1) the total value of the regulated firm's investment in each segment of its business, 2) the total operating costs of the firm, and 3) the level of return on that investment that was 'fair'. Once this information was gathered, the commission would determine a total revenue for the firm equal to its operating costs plus the fair return on its investment. Rates would then be set so as to provide this total revenue. The court mandated that the rates were to be determined assuming that they would have no effect on the quantity of the service that would be demanded by the firm's customers (zero demand elasticity), a rule that was favourable to utilities during that period when rates were falling. Numerous problems had to be overcome before such regulation could be effectively implemented. Although railroads were the first businesses subject to this type of regulation, constant judicial challenges blocked effective implementation for a number of years (Smalley, 1906). Rate regulation that could withstand judicial review apparently did not exist until the creation of regulatory commissions with authority over utilities. The adoption of such commissions occurred on a state-by-state basis, with the majority of states adopting commissions between 1907 and 1914 (Table 1).

The commissions that regulate utility rates operate as quasi-judicial agencies. Each commission has several commissioners, and decisions are determined by vote. Before a rate can be established or changed, a formal hearing is held. Testimony is given under rules of evidence based on accounting data provided for a 'test year'. Commissioners also have power to decide whether a particular utility investment is 'prudent', and can therefore be included in the value of the total investment on which the fair return is calculated. The utilities participate in these hearings, of course, but

Table 1. Year of establishment of state electric regulatory commissions.

Year	Number of states	State names
1887	1	Massachusetts
1905	1	New York
1907	2	Georgia, Wisconsin
1908	1	Vermont
1910	2	Maryland, South Carolina
1911	9	California, Connecticut, Kansas, Nevada, New Hampshire, New Jersey, Ohio, Oregon, Washington
1912	2	Arizona, Rhode Island
1913	10	Colorado, Idaho, Illinois, Indiana, Maine, Missouri, Montana, North Carolina, Oklahoma, West Virginia
1914	2	Pennsylvania, Virginia
1915	2	Alabama, Wyoming
1917	1	Utah
1919	3	Arkansas, North Dakota, Tennessee
1921	1	Louisiana
1934	1	Kentucky
1939	1	Michigan
1941	1	New Mexico
1949	1	Delaware
1951	1	Florida
1963	2	Iowa, Nebraska
1975	3	Minnesota, South Dakota, Texas

Source: Neufeld (2008).

they are also open to essentially anyone wishing to provide testimony. The decision of a commission can be appealed to the judiciary. Commissions generally have full-time staffs reviewing the evidence presented and providing technical support, although their budgets are a small fraction of the administrative budgets of the utilities they regulate.

Regulation prior to state commissions

State rate regulation did not replace open competition among utilities; it replaced regulation by municipalities. Utilities, including electric utilities, must place their distribution systems on land they do not own. As a practical matter, the public rights-of-way must be used for this purpose. This required a 'franchise' from the municipality. These franchises were contracts specifying the rights and responsibilities of both the utilities and the municipalities in which they were located. Franchise terms were fixed but usually long – 20 to 50 years. The terms of the franchises varied considerably among municipalities and often included terms and availability of service, maximum prices, competitive conditions, discounts to the municipal governments, and many other requirements.[2] Because of rapid technological progress, the agreements on maximum pricing usually became non-binding. Unlike commission regulation, there was typically no provision for constant oversight, for altering the terms of the franchise, or for the arbitration of issues resulting from changed conditions.[3] Rarely were utilities offered monopoly protection beyond the first few years of operation. Entry by competing utilities could and did occur.

The period of regulation by municipal franchise was one in which there was widespread concern over corruption in municipal governments. The source of this corruption often involved issues involving municipal utility franchises and arose because the economics of utility service made utilities especially susceptible to holdup. Utilities in general (and electric utilities in particular) had to spend enormous amounts on fixed investment such as generating equipment and distribution networks before they were able to receive any revenue. To be economically viable, a sizable portion (perhaps most) of that revenue had to be dedicated to repaying those fixed costs. By contrast, the utility's operating costs usually necessitated a relatively small portion of the revenue needed to remain in business in the long run. This difference between the revenue needed to cover short-run operating costs and the revenue needed to maintain economic viability created appropriable quasi-rents. If the utility's revenue at least covered operating costs, it had an incentive to remain in operation even if that revenue did not fully cover its capital costs. It did not, however, have an incentive to replace or, perhaps, even to provide adequate maintenance for that capital equipment. Service would deteriorate, and, as the capital equipment completely wore out, would cease. This process, however, could take years, and if corrupt politicians or municipal officials were successful in appropriating those quasi-rents, the results would not be immediately noticeable to residents. The deteriorating service was likely to be blamed on the utility rather than public officials. A documented method by which this appropriation was achieved was through the threat to go into competition with a utility unless a bribe was paid.[4] The threat of this type of corruption would inhibit the development of new utilities, discourage investors from putting funds into the new or growing utilities, and increase the cost to utilities of obtaining the funds required to start and maintain a viable utility (Hausman & Neufeld, 2002). The early utilities generally were not very profitable enterprises (Hausman & Neufeld, 1990).

Two solutions were proposed for the problem of utility quasi-rents and the corruption it induced: municipal ownership and state regulation. Municipal ownership would not have eliminated the opportunity for corruption. It might have created new opportunities through such things as the manipulation of purchasing contracts or the awarding of jobs in return for political favours. It would, however, have made it far more difficult for corrupt officials to appropriate quasi-rents. The municipality itself would have been responsible for service deterioration and also had to depend on outside investors to raise the funds needed for capital expansion.

State regulation might have been ineffective at restraining rates and might have reduced the incentive for utilities to operate efficiently. The requirements imposed by Smith v. Ames, however, gave the quasi-rents of a utility under commission regulation the status of a constitutional right protected by both state and federal courts.

Municipal ownership became important in some states. Prior to 1937 there were three states in which municipally-owned utilities generated most of the power for at least some years and 14 others for which they accounted for at least 20% of total generation. For the nation as a whole, however, municipal utilities, although numerous (often making up a majority of all electric utilities), were small and accounted for less than 10% of either generation or generating capacity during the period when state regulation became widely accepted. The New Deal sparked a resurgence of interest in municipal utilities partly because of concern over the widespread consolidation of private utilities that had occurred through holding companies, especially in the 1920s, and increased dissatisfaction with the efficacy of state rate regulation. In the 1940s, privately-owned electric utilities essentially disappeared in both Tennessee and Nebraska. The political question of whether electricity users were better served by privately-owned (and regulated) utilities or by publicly-owned utilities was a recurrent political issue in the United States. Disagreements over this issue were intensely ideological, emotional, partisan, and characterised by arguments that, at best, contained half truths.

Transmission networks at the adoption of state regulation

The structure of the industry was very different during the period in which most states adopted state commission regulation than it was to become. Electric service was unavailable in rural areas. Transmission was constrained by the maximum voltages then achievable. In 1917, only slightly over 60% of the US population lived in areas where it was even possible to obtain utility-generated electricity. Per capita use of electricity by those living in those areas was only 81 kWh for light and 200 kWh for power (the latter doubtless primarily industrial) (US Bureau of the Census et al., 1920). It is difficult to determine with certainty the extent to which a single utility had an integrated network combining generation, transmission, and distribution. Some evidence from the 1917 special census is shown in Table 2.

Clearly some utilities were distribution-only, depending on other utilities to supply them with power. The number of separate generating stations, however, is only slightly larger than the number of plants, indicating that most of the electric utilities at the time operated with only a single generating station. The number of

Table 2. Character of electric utilities in 1917.

Number of plants	6542
Generating all or part of current	5124
Purchasing all current	1418
Number of separate generating stations	5952
Number of cities, towns, etc. served by all stations	13,716

Source: US Bureau of the Census (1920).

towns is almost twice the number of plants; when a utility served more than one town, it must have often still used only a single generating station. To what extent did transmission networks cover large areas and interconnect multiple generating stations and multiple utilities? At the time regulation was adopted (as early as 1910), Chicago was regarded as having the most advanced electrical supply system in the country, if not the world, and credit for this achievement is accorded to Samuel Insull, an industry leader who was president of Commonwealth Edison, which, by this time, had become the largest monopoly utility serving a large metropolitan area (Hughes, 1983, p. 204). Insull's biographer, Forrest McDonald, describes an experimental project Insull launched in 1910 to extend Chicago's network to some 20 towns located in an adjacent county (McDonald, 1962, pp. 137–145).[5] McDonald makes clear that this action was regarded as 'radical' by most in the industry, although this type of system was quite small by modern standards, including only a single metropolitan region and surrounding area. These networks do not appear to have extended across state boundaries. It seems safe to conclude that at the time state commission regulation was adopted by most states, the typical utility served a single relatively small area from a single generating station in a single state.

An important issue, to which we will return later, is the question of the proper scope of a firm operating in the electric utility industry. The functions of an electric utility have traditionally been conceptually divided into three parts: generation, transmission, and distribution. The issue of firm scope is whether these functions are best handled by a single firm or by separate firms, and this question has become integral to recent deregulation efforts. At the time state commission regulation was adopted, most utilities incorporated all three functions although transmission networks were primitive compared to what they would become, and some utilities performed only the distribution function. State regulation assigned an existing utility a defined service area and protected all the functions performed by the existing utilities from competition. Although changes in the industry's structure did occur, regulation made it very difficult for any change not in the interest of an existing utility. This may be an important reason that firms in the industry remained vertically integrated (generation, transmission, and distribution). This does not mean that this scope was not economically optimal. It does, however, raise the question of the optimal desirable scope since, as technology developed, market forces were significantly hindered from changing that scope.[6]

The benefit of large networks

As technology developed it became possible to conceive of and create ever larger transmission networks interconnecting multiple generating sources and many users over a large geographical area. These networks provide important economic benefits by reducing the amount of generating and other equipment a utility needs. The size (and cost) of much of the capital equipment required to produce and distribute electricity is determined by the peak demand for electricity, the maximum amount of power that must be supplied, not the total amount of energy supplied.[7] For example, consider two different utilities each producing and distributing 24 GWh of electricity per day. Both are thus providing the same amount of electrical energy to their customers daily. One utility's customers use electricity only during a single hour each day, and that utility must provide all 24 GWh during that single hour; at all other hours electricity is not used at all. The other company's customers use electricity at a constant rate; each hour of the day that utility needs to provide one GW of power. The first company must have generating equipment with a power capacity 24 times of that of the second company, and much of the transmission and distribution equipment will also have to be sized to handle the larger power. That same equipment could be used during the other 23 hours of the day; as long as less than 24 GW were required during those other hours, it could be provided at very low cost.[8] It is far more economical to provide electricity when its use does not vary greatly. Different types of users (different industries, residential users, other businesses) will have peak power demands at different times of day. A large network can also benefit from users' peak power demands occurring at different times if it spreads over multiple time zones. Large networks tend to have less fluctuation in power demands than small networks.

Large networks provide other economic advantages. There are different types of generators that can be used to produce electricity. Some of these have high up-front capital costs but very low operating (fuel) costs. Others have much lower capital costs but higher operating costs. A large network can permit the employment of an efficient mix of these different types. Those with high capital costs and low operating costs provide the 'base load'. They operate at full capacity all of the time. The additional amounts of electricity required at certain times can be provided by the generating plants with lower capital but higher operating costs. In this way the total cost of providing electricity can be minimised. In addition, large networks reduce the proportion of generating capacity that must be kept in reserve and used when a generator breaks. Finally, networks are particularly important when hydroelectricity is available. The supply of electricity from a hydroelectric source depends on the amount of water flowing through a dam. This is often subject to seasonal and random fluctuation. Electricity is most valuable when it is available when needed rather than just when rainfall is favourable. The most efficient use of hydroelectricity occurs when other types of generating equipment on the network can make up for variations between the need for electricity and the availability of hydropower. There are also economies to networks that include multiple dams on the same river system. Dams on tributaries can retain water during periods of high availability and release it during periods when rainfall is low. The water that flows through a tributary dam during such periods will also flow through dams on the main river, permitting the same water to generate electricity twice, and enabling greater control over the availability of hydroelectricity (Hausman & Neufeld, 2004).

Economic barriers to large networks

The increased benefit of larger networks with fully integrated transmission facilities face a special impediment originally described by Oliver Williamson (Williamson, 1985). Williamson's objective was to explain why some transactions occur under administrative control between divisions of the same firm while others occur between different firms in the marketplace. He identified specific transaction costs present in markets but absent within firms. Transactions occurred intra-firm when that transaction required long-lived expensive capital equipment with little value if it became unusable as originally intended. The agreement commits one of the parties to spend money on an expensive piece of equipment whose value could not be recovered were the agreement ended prematurely. Once that capital equipment has been put into service, the party who bore those large fixed costs will have sharply lower bargaining power should it become necessary to negotiate any change in the original agreement.[9] Long-term contracts with contingency clauses are designed to handle these situations, but can do so only imperfectly because it is impossible to imagine all future contingencies. The party to the transaction providing the capital faces the risk that the other party will renege. This possibility exists when transactions occur in the market between independent firms. Were the transaction to occur within a firm, the risks associated with and change in commitments would be borne entirely by that firm. There still is a risk that the investment will not prove profitable, but the additional risk faced by one party that the other will engage in opportunistic behaviour is eliminated. Such transactions, therefore, tend to occur within a single firm rather than between different firms.

The capital equipment required for both generation and transmission are very transaction-specific and thus, following Williamson, are unlikely to be provided in response to inter-firm market agreements. The creation or extension of large integrated generation and transmission networks are thus more likely to be supplied by a single firm rather than multiple firms. Regulation likely inhibited the ability of a single company to develop a large integrated network. The types of market forces that have led large retail stores to replace smaller stores were unavailable to the electric utility industry since each existing utility was protected against competition. Consolidation could and did occur, but the monopoly protection would have increased the bargaining power of the utility being acquired versus the acquirer, since the cost of rejecting an offer was simply maintenance of the status quo and a guaranteed regulated profit. Regulation may also have further inhibited agreements between utilities, particularly in different states, to jointly develop integrated networks since, to the risk of a utility reneging on a long-term agreement, was added the risk a commission might order a utility to renege in the interest of that state's residents.[10]

Holding companies

Consolidation became easier through the device of a holding company, an ownership form that became increasingly popular for electric utilities during the 1920s. A holding company would gain control over an operating utility by purchasing that utility's stock on the open market. This did not require the type of negotiation with the utility's management needed with a formal merger, and the holding company did not have to purchase all of the utility's stock in order to gain control. The earliest holding companies were involved in providing financing to their operating subsidiaries.[11] Financing became an increasingly important function of holding companies that would issue their own securities to raise the funds to buy control of operating companies. There were several reasons investors found the securities of holding companies more attractive than those of the operating companies. Holding companies were able to improve the business of their operating companies by achieving the type of consolidation and network expansion mentioned above, by providing technical and management expertise superior to that available to the smaller operating companies, and by economies arising from central purchasing. Holding companies were large, national corporations better known to investors than the myriad local operating companies, and, by combining a number of operating companies, were able to offer investors a degree of diversification; in this sense they were like mutual funds specialising in utilities (US Federal Trade Commission, 1928, pp. 169–173).

The involvement of holding companies in the electric utility industry in the 1920s was associated with a period of unusually rapid gross capital formation (Figure 1).

The 1920s was also the beginning of a period in which increased growth in the use of electricity in manufacturing appears to have had major responsibility for the rapid total factor productivity improvements that occurred in American industry (Schurr, Burwell, Devine, & Sonenblum, 1990, especially Chapter 12) There was also, however, a darker side to the holding company system. The intense financial activities, and the huge profits these produced, were responsible for fraud and financial bubbles by the end of the decade both in the market for the purchase of operating companies by holding companies and in the market for the securities issued by the holding companies (White, 1990; Young, 1965, pp. 32–33). Many holding companies adopted shaky financial structures – such as issuing senior securities (bonds and preferred stock) – to finance the purchase of operating company common stock. This resulted in leverage that magnified high profits but could (and would) also result in losses when operating profits fell. Leveraging was

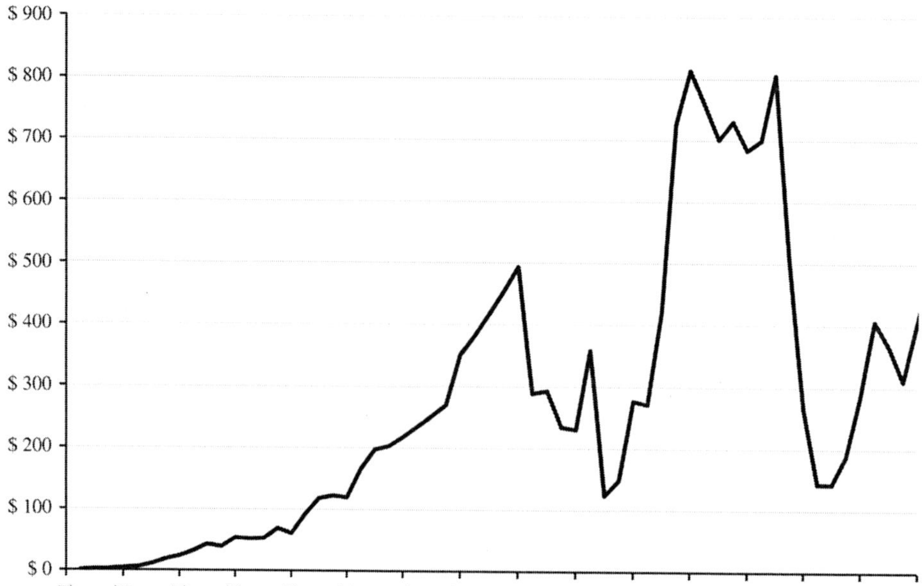

Figure 1. Real gross capital expenditures by electric light and power utilities in millions of 1929 dollars.
Source: Ulmer (1960, pp. 374–375).

amplified through 'pyramiding' whereby layers of different holding companies connected an operating company to a top holding company whose direct investment was a tiny proportion of total assess but nevertheless had control over the entire structure (US Federal Trade Commission, 1935a, pp. 155–166).

Holding companies were able to subvert the system of state commission regulation through several means. The holding companies themselves were not subject to the rules or oversight of regulatory commissions, and they charged fees to their subsidiary operating companies for the various services they performed. These would be presented to regulatory commissions as part of the utility operating expenses which it was entitled to recover from its customers. Regulatory commissions had no ability to investigate the actual costs to the holding companies of providing these services, leaving the holding companies free to improve their profitability by charging excessively large fees to their operating companies. Holding companies were also able to increase the capitalised value of the operating companies, perhaps after themselves paying excessive prices to achieve control. This increased capitalisation supported the issuance of more securities that could increase the operating company's debts. Although regulatory commissions had the power to refuse to increase the value of the utility's property (the 'rate base' whose 'fair value' was the basis for the regulated profits), they were unable to prevent these practices from occurring. Were the regulatory commission to deny the operating utility an increase in the value of its rate base, the result could have been to put the operating company in financial distress, resulting in a deterioration of service (Bonbright & Means, 1932, pp. 167–175; US Federal Trade Commission, 1928, pp. 173–174). As the industry in the late 1920s appeared to be rapidly moving toward a nation-wide monopoly, could state commission regulation effectively protect electricity users from exploitation by a 'power trust'?[12]

Dissatisfaction with this growing concentration of industry control prompted Gifford Pinchot, Governor of Pennsylvania, to sponsor a proposal to radically restructure the industry in ways that presaged aspects of the current deregulatory efforts (Pennsylvania. Giant Power Survey Board, Pinchot, & Ettinger, 1925). Generation, transmission, and distribution were to be provided by separate entities. Negotiations would occur between generators and distributors; a single transmission system would affect the transfer of power from generator to distributor as a common carrier. Fierce opposition by the existing privately-owned utilities helped prevent the plan's adoption.

The apparent impotence of state regulation in the face of holding companies helped stimulate a revival of the debate over whether utilities should be privately or publicly owned. One of the factors stimulating this debate was the growing involvement of the federal government in hydroelectricity (Hausman & Neufeld, 1999). Initially this came from two sources. First, the federal government was constructing dams for irrigation and reclamation purposes, and had begun generating power from one such dam in 1903. Second, the federal government had control over all of the nation's waterways, and its approval was required before any dam could be constructed over such a waterway. In 1920, the Federal Power Act created the Federal Power Commission to streamline the granting of licenses for dam construction. The act contained a 'municipal preference clause'. If a municipality and a private company were both seeking the same license, and their applications were otherwise judged equally attractive, the municipality would be given priority. Concern over the growing power of holding companies led to a study by the

Department of Agriculture in 1916 on the extent to which holding companies had gained control over the electric power industry in general and hydroelectricity in particular (US Department of Agriculture, 1916). In 1925 a Senate resolution directed the Federal Trade Commission to undertake a comprehensive study of holding company control that resulted in reports issued in 1927 and 1928 (US Federal Trade Commission, 1927, 1928). Dissatisfaction in the Senate with this investigation led to a new resolution passed in 1928 to re-investigate holding companies, including their finance and propaganda efforts, with an eye toward the need for federal regulation. The investigation generated a bookshelf of hearings, reports on individual holding companies, and issuance, in 1934 and 1935, of three summary reports by the Federal Trade Commission (US Federal Trade Commission, 1934, 1935a, 1935b).

The investigation showed that private utilities, led by its industry trade group, the National Electric Light Association, had mounted a large and sophisticated propaganda campaign that placed particular emphasis on making the case for private ownership to the press and in schools and universities (US Federal Trade Commission, 1934). Both sides in this debate had considerable support in Congress. During the 1920s, the single issue that most occupied Congress was the question of the disposition of Wilson Dam, an important producer of hydroelectricity originally built by the Army Corps of Engineers to meet anticipated defence needs created by World War I (Neufeld, Hausman, & Rapoport, 1994). Was this valuable resource to be put under public or private ownership and control? A similar controversy arose over the construction of Hoover Dam by the Reclamation Bureau. Herbert Hoover became involved with the issue in 1922 when he was Secretary of Commerce, but it was not until 1928 that Congress authorised its construction. Controversy over power control led to a remarkable compromise on the electricity produced at the dam, reflecting the political division and sensitivities: one power plant was operated by Southern California Edison for the benefit of privately-owned utilities; a different power plant on the other side of the river was operated by the City of Los Angeles for the benefit of publicly-owned utilities. The federal government, which maintained ownership over both facilities, did not directly sell electricity to either. Instead it 'leased' the falling water that flowed through the hydroelectric turbines (Hausman & Neufeld, 1999, pp. 57–60).

The proponents of public ownership in this period had an ultimately important ally in Franklin Roosevelt, the Governor of New York. Across the northern border of New York State lay the Province of Ontario and the region served by Ontario Hydro, a utility owned by the Canadian Province established to exploit Canada's share of the hydroelectricity produced by Niagara Falls. Residents of New York chafed from the fact that electricity rates were far lower in southern Ontario than northern New York. Furthermore, rural residents in Ontario were provided with electric service, while their southern neighbours had to do without. In 1929 Roosevelt had data collected that demonstrated these facts. The issue became an important one for the Democratic Party in the election of 1930 when it brought them support from both upstate Democrats and Republicans (Davis, 1985, pp. 91–92). Shortly after that election, with the support of Roosevelt, the legislature created the New York State Power Authority with functions similar to that of Ontario Hydro. Roosevelt embraced the idea that public ownership of only a portion of the electric utility industry could restrain the behaviour of private utilities in a way that state commission regulation had failed to do. It would function as a type of regulation itself. Publicly-owned utilities would provide a 'yardstick' of rates and service against which private utilities could be judged. Private utilities would have to respond to this yardstick to avoid the 'birch rod' of public takeover resulting from informed consumers who could see the extent to which they were being exploited.[13] In this way privately-owned utilities would be subject to the regulating effect of at least potential competition from publicly-owned utilities.

The New Deal and demise of holding companies

The Great Depression and the New Deal shifted the political landscape in favour of public ownership, but the rancour between the two sides worsened. In 1933 Congress passed the Tennessee Valley Authority (TVA) Act. This new authority was charged with promoting economic development in a poor section of the country by several means, but the hydroelectric development of the Tennessee River became its most visible undertaking and, for the first time, gave a federal agency a major role in the production of electricity.[14] Wilson dam was transferred to the new authority. The Act specified that the TVA was to give preference in the sale of electricity to utilities owned by governments (state, county, municipal) and cooperatives and was to put special emphasis on serving domestic and rural consumers. TVA's interpretation of this made it unwilling to supply power to any privately-owned utility, and communities that were served by a private utility and wanted TVA power would have to convert to municipal ownership.

In 1934, the residents of Tupelo, Mississippi became the first to receive TVA power. The referendum to replace the existing privately-owned utility with one owned by the municipality was a foregone conclusion once TVA announced that the rates charged by the new municipal utility would be half the former rates. This ultimately resulted in a clamour for publicly-owned utility service to replace existing privately-owned service wherever TVA was prepared to supply current. Private utilities felt themselves put at an unfair advantage since they had no access to the low-cost hydroelectricity offered by the municipal utilities. Although it usually made sense for the privately-owned infrastructure to be transferred to the public utility, a municipality that decided to construct a duplicate distribution system would receive a subsidy from the federal Public Works Administration (PWA), a fact that likely had an adverse effect on the negotiating position of the private utilities.[15] A battle also broke out between the TVA and the private companies over rural electrification, which had largely been neglected by the privately-owned utilities. Each side accused the other of aggressively thwarting the efforts of the other. Private companies, for example, were accused of building 'spite lines', transmission lines into rural areas to disrupt the efforts of cooperatives under formation (The Twentieth Century Fund, 1948, pp. 611–612). The president of the holding company controlling most of the utilities in Tennessee, Wendell Willkie, wrote to President Roosevelt warning that conditions amounted to 'practically open warfare', and actual violence seemed like a distinct possibility (Barnard, 1966, pp. 106–107).[16] Opponents of TVA argued that it amounted to unfair government-subsidised competition with private industry, 'creeping socialism'. Proponents pointed to the increased activity by privately-owned utility companies in rural electrification and the falling rates charged by those utilities as evidence that TVA functioned as an effective regulator of the private industry (Welch, 1939).

In 1935 Congress passed the Public Utility Holding Company Act (PUHCA), including the 'death sentence', that forced the breakup of the large holding company systems that had been created during the 1920s and enjoined any company in the future from acquiring more than a very small proportion of the stock of any operating utility. Holding companies that operated fully integrated interstate systems were allowed to continue but were placed under strict regulations. The Act was bitterly but unsuccessfully fought by the privately-owned utility industry.[17] Its passage marked the nadir of that industry's political power. A concern about that act is that it may have denied the nation benefits from increased integration of electricity networks over wider areas. Such networks require unitary control and were most feasible if under unitary ownership, and holding companies had been responsible for the largest such networks then in existence. Originally the bill would have given the Federal Power Commission new authority to order interconnections. The defeat of this provision was due primarily to the vociferous objections of state utility commissions who, in spite of their own deficiencies, jealously opposed any form of federal regulation that they believed might usurp their authority.[18] This failure to enact this provision effectively prevented the structure of American utilities from evolving to one where larger areas came under common ownership. It seems likely that one effect of this was to inhibit the development of larger integrated networks even as technological progress made them more feasible and more economically desirable. It seems unlikely that the power granted the FPC in the original draft of the act would have completely counteracted that effect, but, in any case, we will never know.

The Twentieth Century Fund undertook a comprehensive look at the electric power industry as of 1940 (The Twentieth Century Fund, 1948). The actual work was done during the years 1940 through 1943, but World War II prevented its report from being issued until 1948, by which time public concern over the issues addressed had muted considerably. Among the report's concerns was the difficulty existing institutions had in encouraging and regulating coordinated power operations over areas larger than the service areas of existing utilities. The report expressed dissatisfaction with both the methodology and the nexus of control of the regulation of private utilities, and seems to suggest more direct involvement by government.[19]

Despite these problems, there have been instances of cooperative action among separately-owned utilities, but these varied widely in the extent of the cooperation, and most fell far short of the integration possible with joint ownership. On one extreme was the existence of transmission lines that connected adjoining utilities but used only for spot power transactions. No contractual agreement would exist, and the transaction would occur only when conditions made it advantageous for both sides. At the other extreme were 'tight power pools' consisting of independently-owned utilities participating in a long-term contractual relationship. Such pools could achieve the same level of integration that characterised common ownership, but the conflicting interests of participants often created friction and made these agreements somewhat unstable, often requiring renegotiation (Joskow & Schmalensee, 1983, pp. 59–77).

The battle that had raged over the relative desirability of private vs. public operation of utilities was largely suspended during World War II and never returned to its pre-war intensity. TVA continued to be controversial until Congress granted the TVA the power to finance projects through the issue of bonds and fixed the boundaries within which TVA could supply power. Before being allowed to issue bonds, TVA had to obtain capital funds through Congressional appropriation, which provided its opponents the opportunity to restate their criticisms. Establishing TVA's service boundaries relieved the threat felt by neighbouring private utilities. Prior to this change, TVA was directed to provide power to governments 'within transmission distance' and was required to provide electricity to any publicly-owned utility that constructed transmission lines to a TVA generating facility or to TVA's main transmission network.[20]

Modern crisis

For approximately the next twenty years, the US electric power industry enjoyed a 'golden age' during which electricity usage increased, prices fell, and political controversy over the industry abated. This ended in the mid to late 1970s, when a confluence of events created a 'perfect storm' for the industry and resulted in radical changes in the industry's organisation and regulation. The industry embraced a technology with which it had had relatively little experience – nuclear power – that was highly capital-intensive. For a variety of reasons, these plants took far longer to bring online than originally planned and their costs ballooned to a multiple of those estimated before construction began. All of this occurred during a period of unusually rapid inflation that resulted in exceptionally high nominal interest rates, resulting in both high construction and interest expenses (Figure 2).

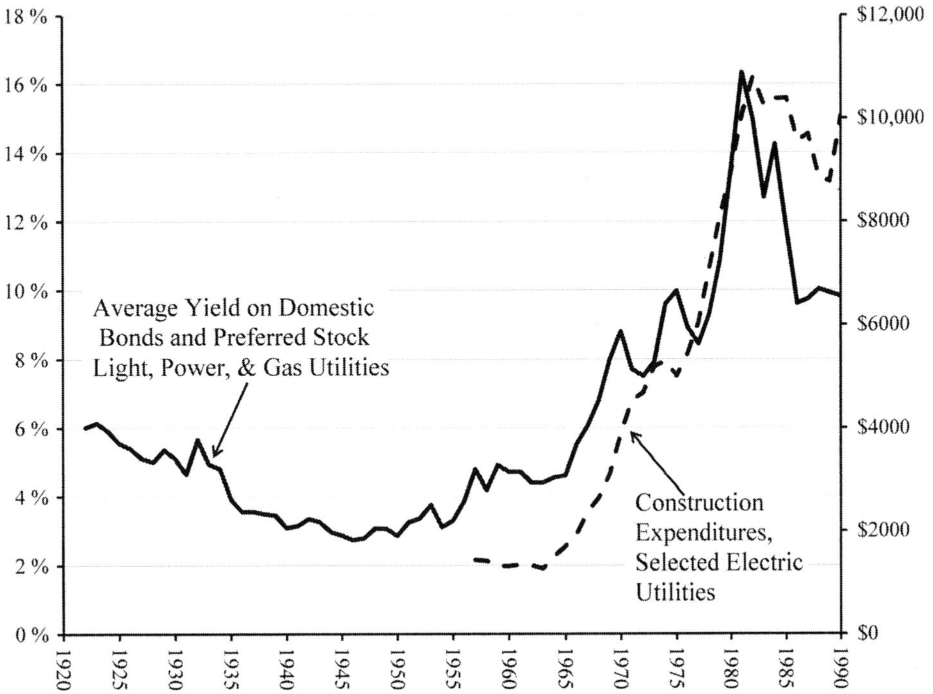

Figure 2. Nominal interest cost (left axis) and construction expenditures in millions of nominal dollars (right axis).
Source: Moody's Investors Service (1991, pp. a2, a15).

In 1973 the Arab oil embargo brought energy issues to national attention. Price controls dampened price increases but resulted in widespread shortages. In addition, the estimated growth in demand that the industry had used for planning turned out to be far higher than actually occurred; the new expensive plants were not needed. Construction was halted on a large numbers of plants on which huge expenses had already been incurred.[21] Beginning in 1971, electricity prices began rising faster than the overall rate of inflation (Figure 3).

The initial proposed policy response to the industry's problems was to reform the rate structure under which electricity was sold. Expenditures on capital were the source of the industry's problem, and the quantity of capital required is a function of peak generation, not total electricity production. Pricing theory supported prices based on marginal cost, the addition to cost of an additional unit of output, rather than the average cost methodology that had arisen under regulation. Under competitive conditions, market forces tend to move prices toward marginal cost. In such markets rate structures are important to producers because they ultimately determine revenue and profitability. Traditional regulation, by contrast, first determined a utility's total revenue and then approved a rate structure that divided that total revenue among the utility's customers. Since rate structures did not determine revenue, utilities had little interest in modifying them, and anachronistic features reflecting conditions that had prevailed decades earlier persisted.[22] Marginal cost principles implied that higher prices should be charged during the system peak rather than during off-peak periods, resulting in proposals for 'peak-load' pricing that, for example, would differentiate the price of electricity by time of day and by season. Marginal cost pricing principles had by then been

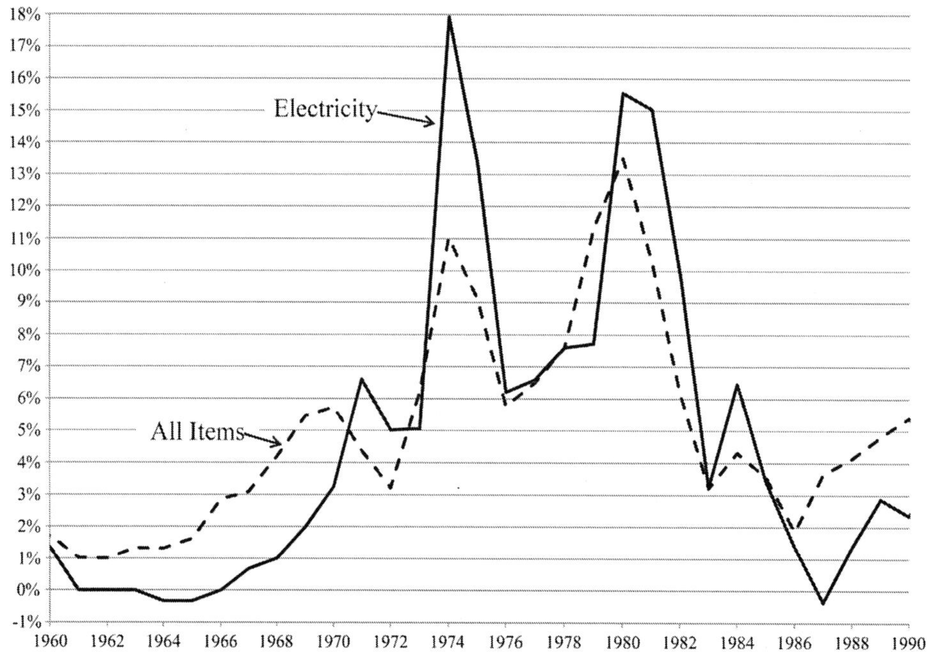

Figure 3. Annual percentage change CPI. All urban consumers.
Source: Carter et al. (2006).

employed in France for a number of years, and had been used to a small extent by some utilities in the US. In 1977, President Carter presented Congress with a National Energy Plan that included requirements that regulatory commissions use marginal-cost methodology to structure rates using nationwide standards, a proposal that would have shifted some of the regulatory nexus from state commissions to the federal government. Unsurprisingly, state regulators were opposed. In addition, opposition to the application of marginal-cost principles came from large industrial users of electricity who enjoyed far lower rates than other users under the traditional rate-making methodology and from much of the utility industry itself.[23] Most of the provisions concerning rate reform were stripped out of the bill.[24]

Genesis of competitive generation

The versions of the plan that ultimately were adopted in 1978 were divided into several different laws. The most important, for our purposes, was the Public Utility Regulatory Policies Act (PURPA), which had a profound effect unexpected at the time it was passed. PURPA created a new class of non-utility electricity generator, 'small power production facility', that produced power either from cogeneration or from renewable resources. No more than 50% of the assets of these facilities could be owned by a regulated utility (Hirsh, 1999, pp. 81–88). Regulated utilities were required to purchase all power offered them by these facilities, and to pay a price based upon the utility's 'avoided cost', to be determined by the state regulatory agency, rather than on the facilities' production costs. Federal regulators were given the authority to exempt these facilities from certain federal regulations otherwise required of electric utilities. Federal regulatory agencies and many state regulatory agencies (such as California's) established prices and rules designed to encourage the development of these non-utility generators, and substantial numbers were created.[25] Some state commissions experimented with bidding systems resulting in competition among these generators to supply electricity to utilities.

PURPA created in the United States a process by which independent power producers were guaranteed access to the utility grid, thereby making it acceptable to think of a utility industry in which most or all power was produced competitively by independently-owned generators feeding a common transmission system from which local distribution systems and large industrial users could obtain electricity. Thus the issue of what was the proper 'scope' of a firm in the electric power industry came into play. Instead of a single integrated firm providing both generation and transmission, generation could instead be provided by a large number of independent firms providing power to a shared transmission network. An important advantage of this structure is that it may reduce the risk of investing in generating equipment because that equipment loses much of its transaction-specific character. If a pre-investment contract with any one electricity user runs into difficulties, the owner of the generating equipment can sell power to any other users connected to the network. The transaction-cost problems arising from failures to anticipate long-run contingencies are not completely eliminated, however.

The idea that electricity could be produced competitively was encouraged by two other concurrent developments. First, all types of regulation, but especially the type of regulation used for electric utilities, in which regulators set prices and controlled entry, were under attack. Such regulation had recently been eliminated for both the airline and trucking industries, some regulations had been removed from the natural gas industry, and significant restructuring allowing competition was occurring in the telephone industry (Vietor, 1994). Second, the disastrous experience with the huge waste of capital that had occurred in the electric power industry made competitive markets seem especially attractive: the risk associated with investment decisions would be borne by the companies making the decisions rather than passed along to consumers.

The changed regulatory climate received further impetus from the Energy Policy Act of 1992. A new class of independent unregulated power producers, 'Exempt Wholesale Generators' (EWGs), was created and freed from rules imposed by PUHCA and PURPA, but made ineligible for the regulator-determined 'avoided cost' rates offered to PURPA facilities. Integrated utilities were allowed to own and operate their own EWGs either within or outside their regulated service areas. The Federal Energy Regulatory Commission (FERC) was charged with developing new rules that would open access to utility-owned transmission systems allowing the EWGs to compete in wholesale markets. Utilities would be required to 'wheel' power from these generators, that is, transmit the power between them and their customers even though the utility owning the transmission network was competing to serve the same customers. The development of EWGs was given a boost by falling natural gas prices, which made it possible for them to quickly (and at comparatively low capital cost) construct gas turbine plants that could produce and sell electricity at very competitive prices, below the prices received by the regulated utilities. As a result, wholesale buyers found it in their interest to contract directly with EWGs rather than the regulated utilities and became active proponents for the introduction of further competition in electricity supply, and industrial users began to advocate 'retail wheeling' whereby they also would be able to contract directly for power.

Struggle over institutional structure

Creating an environment where wholesale and some retail customers would be able to choose among competing generators posed several problems. The most serious was the design of new institutional structures that would somehow make it easy for prospective generating companies to enter the market and serve any user. Although generation might become a competitive industry, no one could envision a method by which transmission could become competitive. All generators would need to supply their power into a common transmission grid from which all users would obtain their power. The operation of an electricity grid is complex and must be under unified control. It would either have to be operated by the government or by a private company subject to regulation. Skilled operation of the transmission grid has always been important; it is necessary to ensure that electricity needs are supplied in the most economical manner, and it is necessary to maintain the greatest possible safety and reliability. With competitive generation it also bears responsibility to ensure that supply is competitive and that all users and all generators have access to the grid. This also requires that, over time, planning and continuing investments be made in the grid in way that maintains and improves system robustness. Separating the control of generation investments from transmission investments makes that planning more difficult. As daunting as these problems were, another problem absorbed even more effort and resources: the problem of 'stranded costs'.

The prices charged by the new class of unregulated generators was often far lower than those charged by the traditional regulated utilities; it was reasonable to assume that once they had the opportunity many users would switch their source of supply to obtain those lower rates. A reason for the high rates charged by traditional utilities was that those rates had to bear the huge costs caused by the cost overruns and the costs of unneeded construction those same utilities had been experiencing over the previous decade or two. A shift by users to the new generators would not reduce those costs; they were sunk costs. Switching generation providers seemed to offer huge savings to electricity users, but those savings exceeded the economic benefits of competitive generation because regulated utility rates had become much greater than marginal costs. The question was who would bear those costs in a regime where customers could abandon the utilities that had incurred them? Perhaps those costs would be borne by the utilities, their shareholders and bondholders. Perhaps they would be borne by those customers who would not or could not change their suppliers, a solution that would have required those electricity users to bear even larger rate increases. Estimates of the total value of stranded costs varied by a factor of 20 – from $20 billion to $200 billion, the latter figure exceeding the value of total industry equity (Hirst & Baxter, 1995). The question of who would bear stranded costs absorbed significant time and resources in all considerations of industry restructuring to the possible detriment of the institution-creation activities that should have been the primary goal.

Discussion of wholesale competition was quickly accompanied by discussions of retail competition. The 'retail wheeling' that industrial users had wanted would be extended to all customers. Under retail competition final consumers would be able to choose the company engaged in such activities as billing. All of these competing companies would, however, use the same distribution system which received its power from a single transmission grid. Electricity is a commodity that is difficult or impossible for competing companies to differentiate, and the activities over which they had control account for a small portion of the retail cost of electricity, leaving little room for price competition.

Although the development of new institutional structures had largely proceeded in a haphazard fashion in the United States, restructuring had progressed much further in other countries. Britain, which had used a number of different institutional structures, privatised its electricity industry in 1990 by dividing it into four classes: generation, transmission, distribution, and supply. Supply companies were to compete for retail customers. The structure of the wholesale generation market became the model for that adopted in the United States. Generators became unregulated competitors. Each morning they would submit bids, containing prices and maximum quantities they were willing to provide at that price (three different bids could be submitted) for each hour on the following day. These bids were used that following day by the transmission company to determine dispatch order; power would be purchased from plants in reverse order of their bid price. The maximum price paid during each half hour period would establish the 'system price', the price paid all generators and the price paid by the distribution companies (to which was also added a transmission charge). Long-term contracts between generation companies and distribution companies would use 'differences'. A distribution company and a generating company that had negotiated a long-term price would reimburse each other for the difference between the contract price and the system price (Newbery & Green, 1996).

The crisis in California

Because of the strong control state regulators had over utilities, the US federal government (via Congressional legislation or FERC), despite being open to the use of competition, could not restructure the industry without the agreement of the states. States' reactions were quite varied. As is so often the case, California led the way (as well as the retreat). California became the first state to warmly embrace both wholesale and retail competition. California's efforts to encourage independent generators under PURPA led, by 1991, to 26% of the state's electricity being generated outside of the state's integrated regulated utilities (Gilbert & Kahn, 1996, p. 204).[26] Debate and discussion of reform of the industry was initiated in the early 1990s by the California Public Utilities Commission (CPUC). At the time, electricity rates in California were about 40% above the national average. In April 1994, the CPUC announced its intention to pursue restructuring and issued what came to be known as the 'Blue Book', a plan for deregulating generation and retail sales that had a galvanising effect on both traditional utilities and on the restructuring movement. 'The major interest groups – the utilities, independent energy companies, large customers, and some important environmental and consumer organisations – all were known to support restructuring' (Blumstein, Friedman, & Green, 2002, p. 18). The state legislature unanimously passed a somewhat modified restructuring plan in August 1996, with implementation to occur on 1 January 1998. Retail choice was instituted. One modification made by the legislature was to reduce and cap retail rates 10% below their existing level (whereas the CPUC had simply called for a rate freeze). In addition, a transition cost was levied on customers to enable the former integrated utilities to recover their stranded costs. Wholesale rates were unregulated. The restructuring plan created two new institutional entities, a Power Exchange (PX) and California Independent System Operator (CAISO), the first to act as a clearing house for daily and hourly markets and to establish prices for a day-ahead market based on bids from market participants, and the second to operate and maintain the transmission grid, with the responsibility to deal with congestion issues (Blumstein et al., 2002, pp. 19–21; US Federal Energy Regulatory Commission, 2005). California's integrated electric utilities were encouraged to divest half of their generating units but went well beyond this, shedding most of their generating capacity by the end of 1999. The time for restructuring was ripe, the rush was on, and by 2001 all but eight states had passed, or were at least considering, restructuring of some form (US Energy Information Administration, 2005). Then another 'perfect storm' hit.

While the new system put in place in California worked smoothly at first, it was not long before troubles began. Due to CAISO's strict and inflexible buying rules, at one point in July 1998 replacement reserve power was priced at almost $10,000 per Megawatt. Purchasers were alarmed. CAISO, with the support of FERC, then adjusted its rules and order returned for a while. But beginning in June 2000 and lasting into the winter of 2000–01 the California electricity market disintegrated, blew up, collapsed. Prices for wholesale electricity tripled and then periodically spiked to unprecedented levels, even in December, with ample available capacity. While generators earned enormous profits, the 'regulated' utilities that were forced to buy in the unregulated wholesale market and sell in the price-capped retail market, suffered badly, as did CAISO. Pacific Gas & Electric, a venerable utility dating to the early days of the industry, declared bankruptcy. CAISO and Southern California Edison were insolvent. The PX simply ceased to operate. Rolling blackouts, causing much inconvenience and economic hardship, occurred throughout the year. In the spring of 2001, the state of California stepped in and became the sole purchaser of electric power (even signing long-term contracts, something prohibited under the restructured market rules), spending over $1 billion per month (Blumstein et al., 2002, p. 27; Borenstein, 2002, p. 191; Faruqui, Chao, Niemeyer, Platt, & Stahlkopf, 2001, pp. 29–52).

In the aftermath of the California debacle, there have been many attempts to explain, or explain away, what happened. It has been blamed on flawed market rules, supply/demand imbalance, the retail price freeze which insulated final customers from what was going on in the market, higher prices for key inputs, outages due to ageing generators, too little transmission capacity, insufficient rain, and/or a bad winter storm. Some prominent economists, as well as FERC, have recognised that market power existed and was exerted brutally. FERC formally found prices to be 'unjust and unreasonable' (Faruqui et al., 2001, p. 45). It is now known that Enron traders (and those of other marketers) manipulated generator dispatch to artificially create the appearance of transmission congestion, the alleviation of which Enron was richly rewarded (until its traders were caught on tape or in memos discussing schemes with names like Death Star, Fat Boy, Get Shorty, Black Widow, Big Foot, Cong Catcher, Forney's Perpetual Loop, Ricochet, and Red Congo) (DeCesaris, Leonard, & Zona, 2005; US Federal Energy Regulatory Commission, 2005).[27] Whatever the cause, or more likely, mix of causes, the crisis had a profound effect around the country. Suddenly, restructuring (with its retail choice and attendant deregulation of some forms of generation) appeared to be a big mistake rather than a panacea. Between 2001 and 2003, electric utility restructuring was suspended in 26 states. Eight states had never begun the process, which left sixteen states that remained actively engaged in restructuring. Since 2003, five of those sixteen states have suspended parts of restructuring or are studying the re-imposition of regulation, leaving eleven states still engaged in restructuring. Many of these states are in the northeast, as well as Illinois, Texas, Arizona, and Oregon (US Energy Information Administration, 2003).[28]

Conclusion

The history of electric utility regulation in the United States was partly determined by the economics of the industry – on the supply side by its cost structure, network characteristics, and lack of storability – on the demand side by its price inelasticity for all but the largest consumers, and partly by politics. These factors influenced the institutions that were created to regulate the industry, a process also complicated greatly by US federalism. Having quickly moved from novelty to essential commodity by the beginning of the twentieth century, politicians had an interest in the performance of the industry. Unlike most other industries in the United States, government at all levels (local, and eventually state and federal) became producers, and both government and private firms coexisted in the same industry.[29] Regulation began at the municipal level (franchise regulation), moved to the state level in the early twentieth century (rate of return regulation), gained an added layer of federal control in the 1920s and 30s (dealing almost solely with interstate transmission of power), and, in some areas changed profoundly during restructuring and deregulation movement of the 1980s and 90s. Although those changes seemed to have considerable momentum, events in a single state at the turn of the millennium bought them to a halt in all states in which they had not already been adopted.

The economics and politics of the electric utility industry differ from those of any other industry, even other network utilities. Stable agreement over the ideal institutional framework within which this vital industry operates has been elusive. Problems periodically arose with existing frameworks that received widespread attention and often created intense debate. Public policies designed to address those problems would have radical effects on the organisation of the US electric utility industry. Those policies, however, either failed to effectively address the problems, or eventually had new and unintended consequences. Sometimes the industry would enjoy a period of quiescence, but eventually widespread attention to then-existing problems would trigger a new round of attempted policy solutions in a continuing cycle.

Notes

1. Despite these variations, the majority of electric power sold in the United States for most of the twentieth century was provided by privately-owned, vertically integrated utilities subject to the control of state regulatory agencies.
2. A two-volume effort to catalogue franchise terms was published in 1910 and 1911 (Wilcox, 1910, 1911).
3. Some of the latest franchises did include such provisions (Priest, 1993).
4. For more extensive discussion of quasi-rents and corruption, and evidence that they led to state regulation, see Neufeld (2008).
5. It is not clear whether Insull gained ownership of the existing utilities or simply enabled them to stop their own generation and become involved only in distribution.
6. Compare the automobile industry. The first modern automobile company, Ford, pursued a policy of extreme vertical integration taking on such things as smelting steel and even growing rubber trees for tires. By contrast, modern automobile companies depend heavily on outside companies for many components.
7. Energy, unlike power, has a time dimension. The gigawatt is a measure of power. The use of one gigawatt (GW) for one hour provides one gigawatt – hour (GWh) of energy. The same amount of energy can be provided by lower power for a longer time period, for example, one half gigawatt of power for two hours.
8. Fuel costs are determined by energy use, but these can be much less than capital costs.
9. This is the same mechanism producing quasi-rents as was responsible for the problem of municipal corruption prior to the adoption of state regulation.
10. In the case of electricity it is also possible for conditions to increase the bargaining power of the owner of a generator resulting in opportunistic behaviour on his or her part. The construction of a generating plant can take years, and this may give the owner of a generating plant significant leverage arising from the threat to withhold the plant's power. Here again, the transaction cost arising from that risk is eliminated if both sides of the transaction are under common ownership.
11. The financing difficulties faced by the early electric utilities caused many equipment manufacturers and technical consulting firms to accept part of the utilities' risk by accepting stock in payment for the equipment or services the utilities needed to operate. These securities were held by subsidiaries that evolved into holding companies (US Federal Trade Commission, 1928, pp. 166–169).
12. In 1929 over 82% of all generation from privately-owned utilities was from ones controlled by holding companies. The three largest holding companies controlled utilities accounting for over 45% of all generation from privately-owned utilities (US Federal Trade Commission, 1935a, p. 38). The position that regulated utilities 'captured' their regulators so that the regulation effectively protected the utilities from competition but did not otherwise constrain their actions to maximise profits may also have affected perceptions over the effectiveness of regulation.

13. For a somewhat later, but clear, exposition of this position, see Olds (1935).
14. Some power production was already occurring on a relatively small scale at dams operated by the Reclamation Bureau. Notable was the Theodore Roosevelt dam on central Arizona's Salt River. By 1923 there were a dozen such projects.
15. Actual duplication of service was rare, and, in 1938, the PWA would provide funds for a duplicate system only if the municipality had tried unsuccessfully to purchase the private utility at a reasonable price (The Twentieth Century Fund, 1948, pp. 387–388, 395–398).
16. Wendell Willkie ran as the Republican candidate against Roosevelt in the 1940 presidential election.
17. The industry's lobbying activities were so intense that the Senate undertook an investigation of those activities before the bill was passed (Senate Special Committee to Investigate Lobbying Activities, 1935a, 1935b).
18. The original text of the bill can be found in (US Senate Committee on Interstate Commerce, 1935). The position of state regulators was made by the chairman of the legislative committee of the National Association of Railroad and Utilities Commissioners, pp. 725–731.
19. For example, the report suggested that main transmission lines be operated by the federal government (The Twentieth Century Fund, 1948, p. 694).
20. Sections 11 and 12 of the Tennessee Valley Authority Act of 1933.
21. Duke Power cancelled one plant on which sunk expenditures amounted to over 30% of the company's pre-cancellation net worth (Hearth, Melicher, & Gurley, 1990, p. 102).
22. On the dominant rate structure for industrial users see Neufeld (1987).
23. Evidence of this can be found in the public statements made during the bill hearings (US Senate Committee on Energy and Natural Resources, 1977). On the position of state regulators see the statements of officials from NARUC, pp. 760–800. On the views of large industrial electricity users see the statement of the executive director of ELCON, an association of 12 of the nation's largest industrial users of electricity (pp. 494–511). Several large industrial electricity users issued booklets or articles attacking marginal cost concepts including Air Products & Chemicals and Kaiser Aluminum (Kaiser Aluminum & Chemical Corporation, 1976; Sherry, 1977).
24. The Department of Energy (DOE) was created in 1977 and the Federal Power Commission was renamed the Federal Energy Regulatory Commission (FERC) with status as an independent regulatory agency within the DOE.
25. Co-generation was the basis of more of these facilities than renewable power. In some cases the prices utilities had to pay for this power was so high that some argued this created 'PURPA machines', companies that, in order to qualify as co-generators, undertook some other production that was not economically justified on its own, but whose losses could be covered through the profits made from selling power at prices above the cost of production (Barclay, Gegax, & Tschirhart, 1989, p. 226).
26. Non-utility generation also accounted for 20% of all generation in Texas and 27% in Louisiana.
27. Decesaris, Leonard, and Zona argue that not all of these strategies were pure market manipulation but that most of them at least gave false information. They argue that FERC's market manipulation rules are difficult to apply. This raises the question: what kind of market is it that requires detailed and complex written rules to work? (DeCesaris et al., 2005, p. 177).
28. Texas has been committed to restructuring. Even though it had relatively low utility rates, the state welcomed freewheeling competition and there are over 100 electricity retailers in the state. The process, however, hit a bump in late May and early June 2008, when, for unknown reasons prices spiked to as high as $4,000 per mWh, and actual retail rates rose to almost 19 cents per kWh. Again, congestion and strategic (but not illegal) bidding drove prices up. It was stated that some generators expected to receive prices in the $2500 mWh range for at least a few hours since they claimed that receiving these occasional prices were needed to survive.
29. Other municipal utilities in the United States (e.g., water, transit) are owned and operated by local governments in some areas and private companies in others.

References

Barclay, P., Gegax, D., & Tschirhart, J. (1989). Industrial cogeneration and regulatory policy. *Journal of Regulatory Economics, 1*, 225–240.

Barnard, E. (1966). *Wendell Willkie: Fighter for freedom*. Marquette: Northern Michigan University Press.

Blumstein, C., Friedman, L.S., & Green, R. (2002). The history of electricity restructuring in California. *Journal of Industry, Competition and Trade, 2*, 9–38.

Bonbright, J.C., & Means, G.C. (1932). *The holding company: Its public significance and its regulation*. New York: McGraw-Hill.

Borenstein, S. (2002). The trouble with electricity markets: Understanding California's restructuring disaster. *Journal of Economic Perspectives, 16*, 191–211.

Carter, S.B., Gartner, S.S., Haines, M.R., Olmstead, A.L., Sutch, R., Wright, G., & Lindert, P.H. (2006). *Historical statistics of the United States millennial edition online*. New York: Cambridge University Press. Retrieved from http://hsus.cambridge.org/

Davis, K.S. (1985). *FDR the New York years 1928–1933*. New York: Random House.

DeCesaris, M., Leonard, G., & Zona, D. (2005). Energy trading strategies in California – market manipulation? In M.A. Crew & M. Spiegel (Eds.), *Obtaining the best from regulation and competition* (pp. 161–178). New York: Kluwer Academic.

Faruqui, A., Chao, H.-p., Niemeyer, V., Platt, J., & Stahlkopf, K. (2001). Analyzing California's power crisis. *The Energy Journal, 22*, 29–52.

Gilbert, R.J., & Kahn, E.P. (1996). Competition and institutional change in US electric power regulation. In R.J. Gilbert & E.P. Kahn (Eds.), *International comparisons of electricity regulation* (pp. 179–230). Cambridge: Cambridge University Press.

Hausman, W.J., & Neufeld, J.L. (1990). The structure and profitability of the US electric utility industry at the turn of the century. *Business History, 32*, 225–243.

Hausman, W.J., & Neufeld, J.L. (1999). Falling water: The origins of direct federal participation in the US electric utility industry, 1902–1933. *Annals of Public and Cooperative Economics, 70*, 49–74.

Hausman, W.J., & Neufeld, J.L. (2002). The market for capital and the origins of state regulation of electric utilities in the United States. *Journal of Economic History, 62*, 1050–1073.

Hausman, W.J., & Neufeld, J.L. (2004). The economics of electricity networks and the evolution of the US electric utility industry, 1882–1935. *Business and Economic History On-Line, 2*. Retrieved from http://www.thebhc.org/publications/BEHonline/2004/HausmanNeufeld.pdf

Hearth, D., Melicher, R.W., & Gurley, D.E.J. (1990). Nuclear power plant cancellations: Sunk costs and utility stock returns. *Quarterly Journal of Business and Economics, 29*, 102–115.

Hirsh, R.F. (1999). *Power loss: The origins of deregulation and restructuring in the American utility system*. Cambridge, MA: MIT Press.

Hirst, E., & Baxter, L. (1995). How stranded will electric utilities be? *Public Utilities Fortnightly, 133*, 30–32.

Hughes, T.P. (1983). *Networks of power: Electrification in Western society*, 1880–1930. Baltimore, MD: Johns Hopkins University Press.

Joskow, P.L., & Schmalensee, R. (1983). *Markets for power: An analysis of electric utility deregulation*. Cambridge, MA: MIT Press.

Kaiser Aluminum & Chemical Corporation. (1976). *At issue: Electricity. Pricing a critical resource in an energy-short environment*. Oakland, CA: Kaiser Aluminum & Chemical Corporation.

McDonald, F. (1962). *Insull*. Chicago: University of Chicago Press.

Moody's Investors Service. (1991). *Moody's public utility manual*. New York: Moody's Investors Service.

Neufeld, J.L. (1987). Price discrimination and the adoption of the electricity demand charge. *The Journal of Economic History, 47*, 693–709.

Neufeld, J.L. (2008). Corruption, quasi-rents, and the regulation of electric utilities. *The Journal of Economic History, 68*, 1059–1097.

Neufeld, J.L., Hausman, W.J., & Rapoport, R.B. (1994). A paradox of voting: Cyclical majorities and the case of muscle shoals. *Political Research Quarterly*, *47*, 423–438.

Newbery, D.M., & Green, R. (1996). Regulation, public ownership and privatisation of the English electric industry. In R.J. Gilbert & E.P. Kahn (Eds.), *International comparisons of electricity regulation* (pp. 25–81). Cambridge: Cambridge University Press.

Olds, L. (1935). Yardsticks and birch rods. *Harper's Magazine*, *171*, 648–659.

Pennsylvania. Giant Power Survey Board., Pinchot, G., & Ettinger, J.B. (1925). *Report of the giant power survey board to the general assembly of the commonwealth of pennsylvania.* Harrisburg, PA: Telegraph Printing Co.

Priest, G.L. (1993). The origins of utility regulation and the 'theories of regulation' debate. *Journal of Law & Economics*, *36*, 289–323.

Schurr, S.H., Burwell, C.C., Devine, W.D., Jr, & Sonenblum, S. (1990). *Electricity in the American economy: Agent of technological progress.* New York: Greenwood Press.

Sherry, E.V. (1977). Cutting the marginalists' Gordian knot. *Public Utilities Fortnightly*, *99*, 21–23.

Smalley, H.S. (1906). Railroad rate control. *Publications of the American Economic Association*, *7*, 327–473.

The Twentieth Century Fund. (1948). *Electric power and government policy: A survey of the relations between the government and the electric power industry.* New York: The Twentieth Century Fund.

Ulmer, M.J. (1960). *Capital in transportation, communications, and public utilities: Its formation and financing.* Princeton, NJ: Princeton University Press.

US Bureau of the Census, Hartley, E.F., Lincoln, E.E., Sandford, F.L., Ladd, S.B., Countryman, W.A., Fuller, G.F., & Freeman, C.A. (1920). *Central electric light and power stations, 1917.* Washington, DC: US Government Printing Office. (Reprinted 2006, New York: Ross)

US Department of Agriculture. (1916). *Electric power development in United States and concentration in ownership and control* (Vols 1–3). Washington, DC: US Government Printing Office.

US Energy Information Administration. (2003). *Status of state electric industry restructuring activity as of February 2003.* Retrieved from http://www.eia.doe.gov/cneaf/electricity/chg_str/restructure.pdf

US Energy Information Administration. (2005). *Electric power industry restructuring fact sheet.* Retrieved from http://www.eia.doe.gov/cneaf/electricity/page/fact_sheets/restructuring.html

US Federal Energy Regulatory Commission. (2005). *The Commission's response to the California electricity crisis and timeline for distribution of refunds.* Retrieved from http://www.ferc.gov/legal/staff-reports/comm-response.pdf

US Federal Trade Commission. (1927). *Control of power companies.* Washington, DC: US Government Printing Office.

US Federal Trade Commission. (1928). *Electric power industry: Supply of electrical equipment and competitive conditions.* Washington, DC: Government Printing Office.

US Federal Trade Commission. (1934). *Summary report: Efforts by associations and agencies of electric and gas utilities to influence public opinion.* Washington, DC: Government Printing Office.

US Federal Trade Commission. (1935a). *Summary report on economic, financial, and corporate phases of holding and operating companies of electric and gas utilities.* Washington, DC: Government Printing Office.

US Federal Trade Commission. (1935b). *Summary report: Survey of state laws and regulations, present extent of federal regulation and the need of federal legislation, conclusions and recommendations, and legal studies in support thereof.* Washington, DC: Government Printing Office.

US Senate Committee on Energy and Natural Resources. (1977). *Public utility rate proposals of President Carter's energy program (part e of s. 1469). Part 1* (27, 28 July, 7 Sept.).
US Senate Committee on Interstate Commerce. (1935). *Public Utility Holding Company Act: Senate Hearings. Part 1*, 74th Cong.1. (16–18, 23–27, 29 April).
US Senate Special Committee to Investigate Lobbying Activities. (1935a). *Investigation of lobbying activities. Part 1*, 74th Cong.1. (12, 16–19, 23 July).
US Senate Special Committee to Investigate Lobbying Activities. (1935b). *Investigation of lobbying activities. Part 2*, 74th Cong.1. (24–26, 29, 30 July, 1, 7 August).
Vietor, R.H.K. (1994). *Contrived competition: Regulation and deregulation in America.* Cambridge, MA: Harvard University Press.
Welch, F.X. (1939). The TVA as a factor in public utility regulation. *The Georgetown Law Journal, 27*, 667–679.
White, E.N. (1990). The stock market boom and crash of 1929 revisited. *The Journal of Economic Perspectives, 4*, 67–83.
Wilcox, D.F. (1910). *Municipal franchises* (Vol. 1). New York: McGraw-Hill.
Wilcox, D.F. (1911). *Municipal franchises* (Vol. 2). New York: McGraw-Hill.
Williamson, O.E. (1985). *The economic institutions of capitalism: Firms, markets, relational contracting.* New York: Free Press.
Young, H.H. (1965). *Forty years of public utility finance.* Charlottesville: University Press of Virginia.

The 3 Rs: Regulation, risk and responsibility in British utilities since 1945

Martin Chick

School of History, Classics and Archaeology, University of Edinburgh, Edinburgh, Scotland

Before privatisation, required rates of return and test discount rates were being applied to utility and other nationalised industries. One effect of this new approach was to promote more marginal-cost based tariffs which could fall particularly heavily on low-income groups. This trend was reinforced by privatisation which, when accompanied by market liberalisation, increased uncertainty about the likely returns on capital investment projects. Both of these issues, the treatment of poverty and coping with uncertainty, were of long-standing concern to the Austrian school of economics. Where Austrian economists differed from liberalising governments was in their locating of responsibility.

Introduction

Both as nationalised and then as privatised industries, utilities such as electricity, gas, water and telecommunications were subject to test and regulated rates of return on capital investment. While in their early years as nationalised industries, the capital performance requirements were vaguely embedded in an instruction to cover costs taking one year with another, from the early 1960s initially required rates of return and later test rates of discount were specified. Increasingly reflecting, in principle at least, the wish to cover long-run marginal costs, this approach towards capital investment affected both the level and structure of pricing tariffs. Whereas the early average-cost pricing of the nationalised industries had effectively cross-subsidised between the differing connection and supply costs of the utility, long-run marginal cost pricing structures attributed costs according to where and by whom they were incurred. This gave rise, especially after privatisation, to concerns that low-income groups were being disadvantaged by pricing structures in which standing charges bulked disproportionately large. Particular concerns were expressed about the consequences for low-income groups of the privatisation and regulation of the electricity and water utilities, and this article will concentrate on these two utilities. Not only did each produce output for which there were no complete substitutes, but as capital-hungry industries both provide loci for a further issue concerning capital investment, namely the provision of adequate capacity for future supply requirements. In electricity in particular, the introduction of competition into the industry and the privatisation of other fuel and power industries, increased the risks of making sunk capital investment in the industry precisely because the returns were less certain than they had been. The problem had shifted in part from securing adequate returns on utility capital investment to offering sufficient returns to solicit further investment. The increased risk arose from the liberalisation of the industry, and as such revived memories of the early-Austrian school of economists and their concern that theoretically static analyses of what were actually dynamic markets understated the importance of risk and, at times, uncertainty in shaping the fixed capital investment decisions. As well as analysing the changing approach to the capital investment of utilities, and its consequences for the pricing and availability of output, this article will also use post-liberalisation anxieties over the future adequacy of capacity as an occasion for revisiting some of the concerns of the Austrian- and neo-liberal economists concerning both capital investment decisions and approaches towards the treatment of low-income groups. In so doing, it will also reflect if, and if so how, there has been a shift in the scope and instruments of government responsibility in ensuring the provision and accessibility of utility output.

Regulation

The initial guidelines on undertaking capital investment given to the newly nationalised industries were vague and implicit in 'covering costs taking one year with another' whose main purpose was to inform pricing rather than capital investment. It seemed to be taken for granted that the costs to be covered would be average rather than marginal, and that this was so was made explicit when the economist James Meade urged the merits of marginal cost pricing on the Lord President, Herbert Morrison. The social ambition of connecting and supplying households cheaply on a cross-subsidised postalisation basis was made clear by Morrison in May 1946:

> The fundamental purpose of nationalisation was ... to secure greater efficiency in the industry ... to reduce costs and provide surplus funds which could be used to extend the benefits of cheap electricity to rural areas where ... electricity was either not available or unreasonably expensive ..... Ultimately, he would like to see a standard charge for electricity throughout the whole country, the tariff varying according to the purposes for which the electricity was being used, and the industry being at liberty to make special arrangements with individual interests ... He realised that this system of postalisation had its opponents among some of the economists, but he did not find the arguments which they advanced convincing. The argument that each consumer should pay the precise cost incurred in bringing electricity to his premises would, if followed to its logical conclusion, prevent even a small municipality from adopting a system of standard charges within its own boundaries.[1]

In fact, the rolling out of the network to distant locations could have been funded separately and without being bundled up in the tariff structure. The concern of economists like Meade was not particularly with the one-off cost of extending the supply system, but with the relationship between current prices and the (short- and) long-run marginal costs in an industry with non-stockable output. From the mid-1950s, against a background of economically spurious but politically charged concern with public expenditure, the Treasury was able to press its case for reformed tariff structures echoing Meade's earlier concern with marginal cost pricing. As explained elsewhere, the Treasury's route to the desired reform of pricing structure was not through pricing itself, but instead by requiring higher returns on capital investment (Chick, 2007, pp. 97–103). The initial break-even requirement gave way to the requirement from 1961 that nationalised industries earn required rates of return, and then, moving from an *ex post* to *ex ante* approach, to requiring proposed capital investment projects to meet a test discount rate set by the Treasury (UK Treasury, 1961, 1967, 1978). This in turn impacted on pricing structures, the bulk supply tariff for electricity moving closer to reflecting the long-run marginal cost of supply (Meek, 1968). The tightening of *ex ante* tests and *ex post* returns on capital investment predated privatisation by more than a decade and continued through privatisation into regulation. While US-style capital regulation was not chosen for fear of regulatory capture and 'gold-plating' over-investment, the RPI-X regulatory mechanism developed by the economists Stephen Littlechild and Michael Beesley, initially for telecommunications, necessarily contained implicit judgements on the permitted returns on capital investment (Beesley & Littlechild, 1989). While technically RPI-X made cross-subsidisation for entry-deterrent purposes possible and of concern, the trend towards pricing at long-run marginal cost continued (Helm & Yarrow, 1988, p. xxvii). This was associated with an interest in segmenting markets so as to reflect specific attributable costs of supply. A similar process had occurred in the United States where regulation moved away from fully distributed cost as the basis for rate-making. Under the fully distributed cost approach, common costs were allocated on the basis of the relative shares of quantities such as output, peak demand, and revenue of attributable cost. From the early 1970s, and using concepts drawn from the theory of co-operative games, an increasingly influential approach was developed by economists, many of them employed by AT&T's subsidiary Bell Laboratories, for allocating responsibility for common costs among services so as to avoid cross-subsidy. The Federal Communications Commission undertook a series of investigations lasting from 1964 until the late 1970s concerning the impacts of different fully distributed cost methods applied to various message, private line and telegraphic services offered by AT&T. One of the many criticisms made of the fully distributed cost approach by economists in giving evidence to the Federal Communications Commission was, to quote William Baumol, that 'fully distributed cost test deals with only one circumstance, that is, it deals with the service as it is operated or has been operated. It involves no incremental comparison. It does not compare the circumstance of the user with the service and in the absence of the service' (Brown & Sibley, 1986, pp. 48–49; Faulhaber, 1975). Pricing at long-run marginal cost did look forwards to what might come rather than backwards to what was, and as such did offer greater encouragement for the marginal systems analysis of future investment. In addition to segmenting markets, one consequence of pricing at long-run marginal cost was to make clearer distinctions between the fixed and operating costs of supply. This was a long way from Morrison's approach to cross-subsidisation and it reflected a changed concern to maximise and reallocate the consumers' surplus. These and similar developments in regulatory tariff-making in electricity in the United States during the 1970s were watched closely by the UK Treasury, in particular transcripts of Alfred Kahn's evidence and analyses at regulatory commissions being sent to Great George Street, London.[2]

Standing charges not only formed a higher proportion of the total bill for low users of output, but, reflecting the emphasis on pricing for capital investment provision, their level was likely to rise faster than marginal unit charges. The combined effect of increased standing and unit charges for electricity fell particularly on older (low usage) and poorer (low income and restricted usage) households (Baker & Blundell, 1991; Burns, Crawford, & Dilnot, 1996, p. 13). In spite of stable production and demand, electricity prices for residential consumers in the UK increased between 1990 and 1995, an increase that was not justified by any change in the cost of oil, gas or coal. Yet while prices to residential consumers rose, those to industrial users in the UK fell by around 25% between 1985 and 1995 (Florio, 2004, p. 226). Discounts were offered in the more competitive markets for large customers (Ernst, 1994).

While standing charges to domestic electricity consumers rose by 47% and unit charges by 40% between 1987 and 1995, compared with a retail price increase of 38%, the average increase was larger in water. In aggregate, the 10 privatised water and sewerage firms in the UK increased the average charge for domestic users between 1989–90 and 1993–94 by 55%, compared to an increase of 39% between 1985–86 and 1989–90. Between particular regions, there was a considerable variation in price changes, by as much as 76% between the two extreme cases. The average bill in the region served by South West rose by 187% between 1989 and 1994, whereas the increase was only 30% for that served by Thames and 24% for that served by North West over the same period. As the regulated price of water rose, so too did the number of families who were unable to pay for a water service. In 1990–91, there were 7673 disconnections of water service in the UK; in 1991–92 there were 21,286; the following year there were 18,636. Whereas in 1989, 1% of families were in arrears on their water bills, by 1994 this figure had risen to 9%. This problem of unpaid bills came about partly because of the gap created between the amount charged by the water companies and a new system of income support instituted in 1988. Previously families dependent on state assistance benefited from a mechanism that automatically paid water bills. Beginning in 1988, families had to pay their water bills using the subsidy provided, which calculated a theoretical charge for water of £1.65 per week, compared with the water companies' effective charge of £2 per week (Florio, 2004, pp. 237–238).

Problems of poverty also coincided with an increase in income inequality. While inequality increased in all OECD countries, it occurred to a much greater extent than average in the UK regardless of whether one considers family income, spending or wages (Florio, 2004, p. 97; Gottschalk & Smeeding, 1997). At the beginning of the 1970s inequality was higher than it had been for at least 30 years. While the rise in inequality which began in the 1970s obviously cannot be attributed to a 'Thatcher' effect, her governments did little to end this reversal of the post-war trend towards greater income equality (Jenkins, 1996, p. 39). While in 1979 the proportion of the British population below 50% of median income was slightly more than 5%, by 1997 it was around 15%. The percentage of the population under 60% of median income doubled over the same period. In 1979, the share of total income held by the tenth decile was 4.5 times that held by the first decile; in 1995 it was 8 times. The gap between those with the highest and lowest incomes widened because the richest decile saw its income grow by 50% in real terms over the period, while the poorest decile suffered a drop in real income of 14% (Florio, 2004, pp. 37, 98; Office of National Statistics, 2000).

The composition of the poorest also changed. In place of the traditional predominance of pensioners in the ranks of the poorest, at the turn of the twenty-first century 70% of the low-income families were unemployed people, single parents, invalids, the infirm, and the 'working poor'. A third of children in the UK lived in families in conditions of poverty, the largest proportion in the European

Union (Florio, 2004, p. 98). Whether one chooses 40%, 50% or 60% of the contemporary average income as a poverty line, there was a major increase in poverty between 1979 and 1992/3. In 1994, 9.9 million individuals in 5.7 million families were dependent on the minimum means-tested benefit Income Support; that is about one person in six in the UK. These figures compare with 4.4 million people in 2.9 million families receiving Supplementary Benefit (Income Support's predecessor) in 1979. By this measure, poverty more than doubled in the decade and a half after 1979 (Johnson, 1996, pp. 11–12).

Without wishing to interfere with pricing structures, the utilities' means to palliate the electricity and water bills of the lowest income groups were limited. To address the problem of disconnection, utilities offered prepayment meters. In 1991 there were 48,000 disconnections for non-payment of the electric bill; this fell to 18,000 in 1992.[3] In both electricity and water, the fall in the number of disconnections was linked to the introduction of prepayment systems: 2.3 million in 1992 for electricity, 780,000 for gas. However, as Florio notes, such a widespread use of prepayment meters disguised the problem of disconnection. Service to households with prepayment meters was not actually interrupted, per se, but if users were not in a position to pay, they were forced into periodic disconnection. According to a study in Bristol and Birmingham conducted by the Rowntree Foundation, the introduction of prepayment meters had a dramatic effect in reducing fuel consumption in the households interviewed. This was achieved for instance by cutting the number of cooked meals and minimising the use of hot water. There is evidence of 'fuel poverty' in the UK after privatisation. Conventionally, the threshold amount of income that can affordably be devoted to energy expenditures is 10%. According to this definition, in 1996 in England alone 5 million households (8.5% of families in the UK) were in a state of fuel poverty (Burns, Crawford, & Dilnot, 1996; Florio, 2004, pp. 238–240). The government was driven to increase cash transfers to particular social groups (e.g. winter fuel payments). Apparently government had not lost the wish to tackle fuel and water poverty, but through privatisation it had effectively given up the means of doing so by acting directly on prices.

Risk

Many of the economic benefits of the privatisation programme were expected to follow on from the introduction of competition in the former public monopolies. To economists, the potential benefits of liberalisation considerably exceeded benefits arising from a transfer of the industry from public to private ownership. The liberalisation of an industry's contestable markets could precede, coincide with or follow the industry's privatisation. In 1982, in the Oil and Gas (Enterprise) Act and in 1983 in the Energy Act, liberalisation was attempted before privatisation. Following the privatisation of British Gas as a monopoly, the task of promoting entry into the industry was left to the regulator. Learning from experience, in the run-up to the privatising Electricity Act of 1989, the government actively sought to restructure the industry and then sell it in its competitive form. Investors fought shy of the nuclear power component of the generating section of the industry with its unknown but probably high decommissioning costs for which inadequate provision had hitherto been made by governments. Privatisation transferred risks about which potential private investors wanted more information. Part of the appeal of competition for its advocates was that information was dispersed throughout the market, that if prices reflected (social) marginal costs then these reflections of resource costs were also transmitted to and fro between consumers and producers, and that the heightened risks of competition provided incentives for efficiency. While by the 1980s economics textbooks could set out clearly the virtues of marginal-cost pricing arising out of perfectly competitive markets, the likelihood of competitive markets leading to efficient resource allocation had been fiercely contested for much of the twentieth century. Meade's early contributions on the use of marginal cost pricing in socialised industries arose in response to his reading and reviewing of Abba Lerner's *Economics of control*. Around the same time as Lerner's book, Oskar Lange offered his analysis of a possible use of centrally planned announcements of prices for resources and the incorporation of these prices by socialist managers in their production plans. As resource shortages or surpluses became apparent, so announced prices would be adjusted (Lange, 1935; Lange, 1937; Lerner, 1936; Lerner, 1937; Kirzner, 1997, p. 15).

Some of the fiercest criticisms of socialist planning predated the work of Lange and Lerner and centred on the issue of information. Interwar criticism from the Austrian school of economists was probably best encapsulated in Ludwig von Mises's 1922 book *Socialism: An economic and sociological analysis*, von Mises's thinking drawing in turn on that of Carl Menger and his *Grundsätze der Volkswirtschaftslehre* (Principles of Economics) as well as on that of Böhm-Bawerk (Böhm-Bawerk, 1891). In his turn, von Mises was to influence Friedrich von Hayek, whose work enjoyed belated popularity with sections of the Conservative party and governments from the mid-1970s. It was in part in response to the interwar debate on the possibility of the socialist economic calculation that von Mises and then von Hayek came to develop an 'Austrian' approach to pricing which differed from that emerging from a combination of the work of Marshall and Walras. One of the central questions asked by the Austrian economists was how prices and markets were formed; that while equilibrium economics might be able to explain how an existing market worked, it could not provide a convincing explanation of how the market came to exist in the first place. As such, it was unlikely to be able to explain how future markets might develop over time, the concept of time being central to the distinction between static and dynamic economic analysis. Much of this thinking was drawn together by Mises in his publication in 1940 of *Nationalökonomie*, later translated and published as *Human action*. Hayek's papers written in response to the socialist calculation debate were gathered together and republished in his 1948 *Individualism and economic order* (Hayek, 1949). While there were significant differences in the ways in which Mises and Hayek respectively identified the essence of their understanding of the theory of price, as distinct from that of mainstream theory, nevertheless their shared scepticism as to what a static view of markets had to offer formed a strong common bond. In an observation predating by over half-a-century later criticisms of the efficient market hypothesis, Hayek observed that: 'it is generally made to appear as if these questions of how the equilibrium comes about were solved. But, if we look closer, it soon becomes evident that these apparent demonstrations amount to no more than the apparent proof of what is already assumed. The device generally adopted for this purpose is the assumption of a perfect market where every event becomes known instantaneously to every member' (Kirzner, 1997, pp. 16–18). For the Austrians a key role for an entrepreneur was to create new information; a manager might exploit existing information that was imperfectly known.

One aspect of Austrian economics developed by Mises and Hayek, and then later by Israel Kirzner, concerned uncertainty about the future, particularly as it affected capital investment decisions. In emphasising concern with uncertainty, the Austrians shared some of Keynes's concerns and the Austrian concern with time and the continual process of forming markets was of interest to John Hicks (O'Driscoll & Rizzo, 1985, Chapter 5). In addition to developing the IS-LM model of Keynes's *General theory*, Hicks wrote three books on capital theory, referring back to the Austrian economists in particular in the last of the three books, that on *Capital and time* (Hicks, 1973, p. 12).[4] In the context of the liberalisation of former nationalised monopolies, the uncertainty affecting capital investment decisions was particularly pertinent given the actual and greater potential exposure of industries like electricity to competition, both inside and from outside the industry. Both in the Victorian and Edwardian system of concessions, then as municipal enterprises and subsequently as nationalised monopolies, assurances on probable returns were given and were frequently accompanied and secured by a period free from competition. The 1870 Tramways Act inaugurated the system of providing limited period franchises which was initially usually for 21 years but was subsequently extended to 42 years in 1888 (Byatt, 1979, p. 8). Nationalisation represented an extension of this process in which (often national) monopolistic conditions were established as the norm, public ownership being preferred to regulation as a means of extracting for consumers the benefits of improvements in productivity. Thus, where privatisation was eventually successful in not only transferring ownership but also in introducing competition into sections of each industry, then it reintroduced risk and uncertainty into markets from which it had been absent from decades. In theory, privatisation also increased each industry's responsibility for the consequences of its own decisions, reduced the role and responsibility of the state, and increased the exposure of industries to economic change. Now, if poor decisions were made, as with the early civil nuclear power programme of the 1950s, then ultimately the state should no longer absorb the loss. If relative competitive prices threatened to alter markets significantly from what had been expected, as had happened with the commercial exploitation of North Sea gas, then the State ought no longer to use a mix of taxation and price interference to mitigate the pace and size of the shift in relative prices. If a nationalised industry like coal mining struggled to find markets in the face of falling oil prices, then the state should not again force other nationalised industries like electricity to increase their coal-burn while at the same time increasing taxes and duties so as to restrict oil and coal imports (Chick, 2006).

Much of the early discussion of the liberalisation of former nationalised monopoly markets focussed on how to encourage new entrants into the industry. As such, a keen concern was to prevent the incumbents from exploiting the combination of existing sunk investments and proportionately low marginal costs to deter entry. What was discussed less was how the increase in risk and uncertainty arising from increased competition in markets would impact on the ability to attract new sunk investment. Indeed, rather than under-investment, it was concern with Averch-Johnson style over-investment which was of most concern to those devising new regulatory incentive structures (Helm & Yarrow, 1988, p. xxi). The mainly RPI-X regulatory periods were kept short to around 5 years, not least because their extension to 10 or 15 years was feared likely to allow sufficient time for a gap to emerge between operating costs and prices during which costs savings would not be passed on to consumers. While nominally it was prices which were being regulated, implicitly price regulation was a form of return on capital regulation. Similarly, the prices which emerged from US-style capital regulation had to be politically acceptable. In the United States, for much of the 1950s and 1960s investor-owned utilities had enjoyed an implicit understanding with regulators that allowed returns on reasonable capital investment would be fair, adequate and pretty predictable. Yet, even with the partial liberalisation of the electricity industry, the risks arising from competition both within the industry itself and from outside, from the oil and gas industries, were sufficiently high to deter large, long-term sunk investments such as in nuclear power stations. The concerns with the ability of competitive markets to draw forth such capital investment which had found expression in the Austrian economists' preference for dynamic over static analysis acquired fresh pertinence. So too did the work of an industrial economist like G.B. Richardson who, as Stigler urged the strengths of the theory of perfect competition in the later 1950s, reflected in 1960 in his book *Information and investment* on the failure of economists, with the notable exception of Hayek, to pay sufficient attention to the type of information required for capital-investment decision-making in and by an industry (Hayek, 1937). Richardson reminded his fellow economists of Menger's pioneering work on the marketability of assets (Menger, 1981; Richardson, 1960, p. 157).[5] Ideas which resurfaced in Keynes's exposition of his concept of liquidity preference could also be applied to investment decisions. Just as individual liquidity preference might reflect individual anxieties about the future, then so too might industrialists prefer various forms of liquidity reflecting the differing marketability of assets to a decision to engage in sunk capital investment. Again, this pointed to a market that was always on the move, and that it ever found equilibrium was debatable. Certainly the founders of modern equilibrium analysis were aware of its transitory delicacy and Walras preferred to approach it, as though playing 'What's the time Mr Wolf?' through a process of *tâtonnements* (Walras, 1954, pp. 169–172, 243–254, 284–295).

In the initial Pool trading arrangements designed by the Austrian-economics-influenced Stephen Littlechild for the newly privatised electricity supply industry, specific payments were made to companies for making existing capacity available and the high system marginal price resulting from the day-ahead auctions did make future investment in nuclear power potentially attractive. However, this high system marginal price arose in part from the ability of two dominant players, National Power and Powergen, to 'game' the day-ahead bidding system. In 1998 System Marginal Price in the Pool was still set more than two-thirds of the time by the two largest generators, and almost all the time by only four generators. Pool prices were still some 10–20% above the new entry level based on the latest gas-fired plant (Littlechild, 2000, p. 27). In time, in response to complaints from large customers and competitors of the electricity supply industry, the financial bids in the Pool gave way to the physical bilateral contracting of the New Electricity Trading Arrangements (NETA). Competition was increased in the industry, not least as National Power and PowerGen agreed to divest a total of 6GW of existing coal-fired plant, which was purchased by Eastern Electricity and subsequently run at higher output than it had previously been. Nine years on from privatisation, the duopoly was only half its size at vesting. In 1998/99 National Power and PowerGen accounted for below 40% of generated output, and this market share was to fall further. Increased competition in and outside the industry, regulatory periods significantly shorter than the lifetime of large sunk investment and the ending of Pool prices offering a considerable margin above the lowest marginal costs of nuclear generators, made it more attractive to earn returns reasonably quickly on new CCGT technology than to place undue trust in the future and sink money into nuclear (Littlechild, 2000, p. 26). The inability and/or unwillingness of government to provide assurances to would-be investors in nuclear power of the level and security of their future returns on their capital investment was consistent with a wish to develop a more competitive electricity market; it did however leave unanswered the questions as to how, if government decided that new large long-term sunk investment was required, it was to be made.

In contrast to the privatisation of the electricity supply industry, that of the water industry retained a series of local monopolies within the industry and provided a system of price regulation which specifically recognised the increasing costs and future capital investment requirements of the industry. Rather than the usual RPI-X form which restricted prices to X percent below the RPI for a specified number of years, the water companies were offered RPI + K where K was an allowable price increase above inflation to be used to finance the investment plans necessary to upgrade capacity and meet quality standards (Hunt & Lynk, 1995, p. 374). At privatisation, the parlous condition of the water industry and the need for new investment to stop leaks and to meet new environmental standards had been apparent. Under the pre-privatisation structure of the water industry established by the 1973 Water Act, 10 RWAs (Regional Water Authorities) had been established and charged with planning and controlling the use of water in each river catchment area. Each RWA was responsible for more than one river basin, and collectively it was hoped to achieve an integrated river-basin management. RWAs were responsible for a range of water uses within their broad remit of supplying of water, disposing of sewage and managing sewerage, and overlaid on these were responsibilities for water resource planning, pollution, fisheries, flood protection and land drainage, water recreation and environmental conservation. By the time of privatisation it was clear that investment by the RWAs had been inadequate, and there was particular concern with the decline in river water quality. At privatisation the activities of production and regulation were separated. The newly-created water supply PLCs lost their environmental activities which were transferred to an external body, the National Rivers Authority. The Drinking Water Inspectorate took over responsibility for household water quality and for the implementation of EC directives, and the economic regulation of the industry was delegated to the newly-created Office of Water Services (OFWAT) (Lynk, 1993). While the functions of supply, regulating and environmentally protecting water supply and sources were separated out, the fundamental economic characteristics of the industry restricted the competitive threats faced by each water company. While the regulator and the regulatory formula encouraged the industry to undertake investment and to price at long-run marginal cost, market structure made this more likely to happen in the water than in the electricity supply industry.

Responsibility

Inasmuch as problems concerning fuel and water poverty and the securing of heavy sunk investment were at worst caused and at least exacerbated by the privatisation of utilities and the liberalisation of their markets, the question arose as to whether addressing such problems ultimately remained, despite privatisation, a responsibility of the state. For Hayek and other neo-liberals, poverty was an outcome of a market process and not necessarily reflective of individual fault. In a market what made an individual unique was 'not his generic but his concrete knowledge, his knowledge of particular circumstances and conditions' since 'it is of the essence of a free society that a man's value and remuneration depends not on his capacity in the abstract but on success in turning it into concrete service which is useful to others who can reciprocate' (Hayek, 1960, p. 81; Plant, 2010, pp. 72–73). While a market was a system for providing and co-ordinating information, there was no presumption that a market produced a 'just price'. As markets produced poverty as an outcome of its process of working, then so too should poverty be addressed, for while poverty and 'deprivation are evils ... they are not injustices' (Acton & Shearmur, 1993; Plant, 2010, p. 16).

Dealing with poverty as an outcome was fundamentally different from making the alleviation and eradication of poverty an aim of policy. Indeed, for neo-liberals the state should not pursue such aims; the state should be nomocratic rather than teleological. Although not himself a neo-liberal, Michael Oakeshott's distinction between a teleocracy (an order devoted to the pursuit of some overall end, goal, or purpose) and a nomocracy (a rule governed order not devoted to the attainment of particular ends) had been developed by Hayek in the second volume of *Law, legislation and liberty: The mirage of social justice* (Hayek, 1976; Plant, 2010, p. 16). In a nomocracy political institutions provide a framework of general rules which facilitate the pursuit of private ends, however divergent such ends may be. It was neither the function of political institutions to realise some common goal, good or purpose nor to galvanise society around the achievement of such a purpose. This distinction between nomocracy and teleocracy had some overlap between the distinction later urged by Friedman between positive and normative economics as well as with the praxeological, rather than jurisdictional or political science, approach of Mises (Friedman, 1953; Plant, 2010, pp. 58–59).[6] Like Buchanan, Mises saw the ends of human action as being subjective and he rejected the idea of all human action being directed at some kind of overall goal, happiness or welfare or whatever. Rather human action was undertaken to remedy some subjectively perceived unease in life (Mises, 2007, p. 13; Plant, 2010, p. 58).

So, while the nomocratic state could address poverty which was an outcome of market operations, it should certainly not seek to adjust the relative outcomes by pursuing egalitarian policies. The 'mirage of social justice' in which a normative preferred distribution was pursued was unappealing to neo-liberals, both in itself and in being teleocractic. To neo-liberals, justice was not about the patterns of distributive outcomes but about the process of the market (Plant, 2010, p. 89). The neo-liberal was concerned with the absolute position of the worst off, not with their relative position. However, to return to earth, it was not entirely clear how given the neo-liberal acceptance of the need to address poverty, how the practical problems of fuel and water poverty were to be alleviated.

That the lowest income groups spent a higher proportion of their household budget on energy than did higher income households was clear from Engel curve data and from Family Expenditure Survey (FES) snapshot evidence (Burns, Crawford, & Dilnot, 1996, p. 10). In 1984 the poorest households spent over 13% of household expenditure on energy while for households with incomes in excess of £17,500 the proportion of total expenditure on energy was around 7%. Yet, in contrast to expenditure on food which fell as a share of household expenditure as income rose, energy expenditure remained fairly constant as a share of household budgets as income increased. Energy could then be viewed both as an absolute and as a participation merit good. Low-income groups were also likely to convert fuel less efficiently into power and heat than were higher-income households. In the mid-1980s, while those with household incomes of less than £3500 per annum were less than three times as likely to have gas central heating as electric, those with incomes in excess of £17,500 p.a. were fourteen times as likely to have gas as electric heating. Income issues aside, in the mid-1980s the 30% of households living in council accommodation and the further 8% living in privately-rented accommodation also had little, if any, choice over the type of heating in their households (Dilnot & Helm, 1987, pp. 34, 36, 38, 45). Being less likely to have gas central heating, the poorest also missed out on the slower increase in gas than in electricity charges (Burns, Crawford, & Dilnot, 1996, p. 13).

The nationalised industries' early practice of pricing at average cost offered cross-subsidised prices to all income groups. As an approach to making essential output available to the poor this was effective, but as a means of welfare provision it was inefficient. As James Meade had argued in 1946, poverty was an income problem and should be dealt with as such. By the 1980s with the trend towards pricing at long-run marginal cost continuing, there was little appetite among policy makers for approaches towards tackling fuel poverty which involved a return to the past practices of interfering with the level or structure of prices. As an income problem, fuel poverty was properly dealt with through social security. However, as the system of social security developed from the 1980s it seemed reluctant to deal with fuel poverty head-on as an income problem. Rather it seemed to target particular groups, such as pensioners, for help or to identify particular circumstances, such as cold weather, in which additional payments would be made. That pensioners spent proportionately more on fuel than non-pensioner households was clear (Burns, Crawford, & Dilnot, 1996, p. 10). While energy expenditure formed on average between 5–7% of the budget of non-retired households, it formed 17% of the budget of single retired households mainly dependent on the state pension, and 12% for retired couples mainly dependent on the state pension. Cold weather in the winter of 1985–86 led to the introduction of the 'trigger temperature' system for single payments in exceptionally severe weather. Neither of these payments was income-related. While in particular circumstances weekly additions to supplementary benefit could be made to cover energy costs, with the replacement of supplementary benefit by income support replaced as part of the social security reforms of 1988, no provision was made for the continuance of such payments. The contrast was with the payment of means-tested housing benefit, which was received by slightly more than 7 million households in the UK, more than one-third of the population. Regional variation in rent had long occasioned additional payments to social security recipients, but nonetheless in the mid-1980s 13% of the expenditure of the poorest group was spent on energy, more than the 10% spent on housing (Dilnot & Helm, 1987, pp. 34, 36, 40–41).

Conclusion

Moves towards pricing utility output at long-run marginal cost began before and continued after privatisation. The concern with pricing was an expression of deeper if episodic political concerns with public expenditure and of an underlying rising drone of interest among economists that the resource costs of capital investment be better reflected in the level and structure of output prices. When prices were adjusted, both their level and their structure impacted on the poorest groups in society whose composition was changing. Rather than groups like pensioners sitting at the bottom of the income ladder, this place was increasingly occupied by unskilled people where no-one in the household worked. The social security reforms effected from the late 1980s sought to increase the incentives to work, and as such seemed reluctant to make equivalent provision for the unemployed as for the old on matters such as assistance with fuel costs. While government did not abandon its responsibility to address the problem of poverty, other policies such as privatisation and social security reform reduced its ability to do so. While Mrs Thatcher alluded to the work of Hayek, her governments' policies were often at times less compassionate and forward-looking than the work of the Austrian and neo-liberal economists. For Menger, von Mises and Hayek, risk and uncertainty were dominant concerns. The inability of markets to fully include the factor of time and to overcome uncertainty was clearly recognised by the Austrian school of economists; hence the urging of a dynamic rather than a static view of markets. The liberalisation of markets in which large sunk investments were required simply increased uncertainty and risk, and made those investments less likely to occur. Increased competition could make the existing assets work harder, but the increased uncertainty about future returns and relative prices pushed time-related discount rates higher thereby reducing investment in new large sunk assets. This might not matter, except that in industries like energy, governments express concern about the security implications, be they environmental, economic or military, of a failure to solicit sufficient long-term sunk investment. While issues of national security are a legitimate area for government activity, most of those concerns, be they with global warming or electricity capacity shortages, currently work backwards to the present from a future projection of danger. Government appears to have shifted its area of responsibilities forward in time, while working with present markets which accommodate time inadequately. It might be better for government to accept responsibility for the present precisely by acting to reduce current uncertainty about the future.

Notes

1. The National Archives, Kew, London. CAB 21/2208, Lord President's Office, Extract from minutes of (S.I.M.)(46) 9th meeting, 22 May 1946.
2. For reasons of space I have omitted discussion of Ramsey pricing. To an extent, Ramsey pricing is a response to financial break-even requirements in which price inelasticity is exploited, rather than a direct response to future capacity resource requirements in current output prices.
3. In 1991 there were 19,000 disconnections for non-payment of the gas bill and 16,000 in 1992.
4. Hicks distinguished between Menger's theory as one of economics *in* time and Böhm-Bawerk's as an economics *of* time (Hicks, 1976).
5. Richardson sees Menger's chapter on The Theory of Money in his *Principles of economics* as anticipating some of the much later analysis provided by J.M. Keynes in *The general theory of employment, interest and money*.
6. Praxeology is the science of the means used in action to achieve some subjectively determined end. It is not the science of ends and cannot be a normative science as ends are determined by individual subjective choice.

References

Acton, H.B., Gordon, D., & Shearmur, J. (Eds.). (1993). *The morals of markets*. Indianapolis, IN: The Liberty Fund.

Baker, P., & Blundell, R. (1991). The micro-econometric approach to modelling energy demand. *Oxford Review of Economic Policy*, *7*, 54–76.

Beesley, M.E., & Littlechild, S.C. (1989). The regulation of privatized monopolies in the United Kingdom. *RAND Journal of Economics*, *20*, 454–472.

Böhm-Bawerk, E. (1891). *The positive theory of capital*. London: Macmillan.

Brown, S., & Sibley, D. (1986). *The theory of public utility pricing*. Cambridge: Cambridge University Press.

Burns, P., Crawford, I., & Dilnot, A. Regulation and redistribution in utilities. *Fiscal Studies*, *16*, 1–22.

Byatt, I. (1979). *The British electrical industry*, 1875–1914. Oxford: Clarendon Press.

Chick, M. (2006). The marginalist approach and the making of fuel policy in France and Britain, 1945-72. *Economic History Review*, *59*, 143–167.

Chick, M. (2007). *Electricity and energy policy in Britain, France and the United States since 1945*. Cheltenham, UK: Edward Elgar.

Dilnot, A., & Helm, D. (1987). Energy policy, merit goods and social security. *Fiscal Studies*, *8*, 29–48.

Ernst, J. (1994). *Whose utility? The social impact of public utility privatisation and regulation in Britain*. Buckingham, UK: Open University Press.

Faulhaber, G.R. (1975). Cross subsidisation: Pricing in public enterprises. *The American Economic Review*, *65*, 966–977.

Florio, M. (2004). *The great divestiture: Evaluating the welfare impact of the British privatisations 1979–1997*. Cambridge, MA: MIT Press.

Friedman, M. (1953). The methodology of positive economics. In M. Friedman, *Essays in positive economics* (pp. 3–43). Chicago, IL: University of Chicago Press.

Gottschalk, P., & Smeeding, T. (1997). Cross-national comparisons of earnings and income inequality. *Journal of Economic Literature*, *35*, 633–687.

Hayek, F.A. (1937). Economics and knowledge. *Economica*, *4*, 33–54.

Hayek, F.A. (1949). *Individualism and economic order*. London: Routledge.

Hayek, F.A. (1960). *The constitution of liberty*. London: Routledge & Kegan Paul.

Hayek, F.A. (1976). *Law, legislation and liberty. Vol. 2: The mirage of social justice*. London: Routledge & Kegan Paul.

Helm, D., & Yarrow, G. (1988). The assessment: The regulation of utilities. *Oxford Review of Economic Policy*, *4*(2), i–xxxi.

Hicks, J. (1973). *Capital and time: A neo-Austrian theory*. Oxford: Clarendon Press.

Hicks, J. (1976). Some questions of time in economics. In A.M. Tang, F.M. Westfield, & J.S Worley (Eds.), *Evolution, welfare and time in economics* (pp. 135–152). Lexington, KY: Lexington Books.

Hunt, L.C., & Lynk, E.L. (1995). Privatisation and efficiency in the UK water industry: An empirical analysis. *Oxford Bulletin of Economics and Statistics*, *57*, 371–388.

Jenkins, S. (1996). Recent trends in the UK income distribution: What happened and why? *Oxford Review of Economic Policy*, *12*, 29–46.

Johnson, P. (1996). The assessment: Inequality. *Oxford Review of Economic Policy*, *12*, 1–14.

Keynes, J.M. (1936). *The general theory of employment, interest and money*. London: Macmillan.

Kirzner, I. (1997). *How markets work: Disequilibrium, entrepreneurship and discovery*. London: Institute of Economic Affairs.

Lange, O. (1935). Marxian economics and modern economic theory. *The Review of Economic Studies*, *2*, 189–201.

Lange, O. (1937). On the economic theory of socialism: Part two. *The Review of Economic Studies*, *4*, 123–142.

Lerner, A.P. (1936). A note on socialist economics. *The Review of Economic Studies*, *4*, 72–76.

Lerner, A.P. (1937). Statics and dynamics in socialist economies. *The Economic Journal*, *47*, 253–270.
Littlechild, S. (2000). *Privatisation, competition and regulation*. London: Institute of Economic Affairs.
Lynk, E. (1993). Privatisation, joint production and the comparative efficiencies of private and public ownership: The UK water case. *Fiscal Studies*, *14*, 98–116.
Meek, R.L. (1968). The new bulk supply tariff for electricity. *The Economic Journal*, *78*, 43–66.
Menger, C. (1981). *Principles of economics* (J. Dingwall & B.F. Hoselitz, Trans.). New York: New York University Press. (Original work published 1871)
Mises, L. von. (2007). *Human action: A treatise on economics* (Vol. 1). Indianapolis, IN: Liberty Fund.
O'Driscoll, G.P. Jr, & Rizzo, M.J. (1985). *The economics of time and ignorance*. Oxford: Basil Blackwell.
Office of National Statistics. (2000). *Social trends*. 30. London: HMSO.
Plant, R. (2010). *The neo-liberal state*. Oxford: Oxford University Press.
Richardson, G.B. (1960). *Information and investment: A study in the working of the competitive economy*. Oxford: Oxford University Press.
UK Treasury. (1961). *Financial and economic obligations of the nationalised industries*. Cmnd. 1337. London: HMSO.
UK Treasury. (1967). *Nationalised industries: A review of economic and financial obligations*. Cmnd. 3437. London: HMSO.
UK Treasury. (1978). *The nationalised industries*. Cmnd. 7131. London: HMSO.
Walras, L. (1954). *Elements of pure economics* (W. Jaffé, Trans. and Ed.). London: George Allen & Unwin.

Regulation redefined: the case of the electrical industry in Brazil and India from the 1990s to today

Pierre Lanthier

Introduction

The growth of electricity production in emerging countries has been quite impressive in recent years. India is now the fifth biggest producer in the world while Brazil ranks tenth. Nevertheless, the growth of demand is even stronger, thanks to industrial and urban development and also to the advent of a significant middle class. Shortages are quite common in these countries. It is necessary to increase the electricity output substantially and consequently to attract investment, both local and foreign.

The dominant theories during the 1970s and 1980s concerning how public utilities could be regulated had one major purpose: to attract private (and, especially, international) investment. To achieve this, regulators and governments set three goals: to put an end to 'natural monopolies', deemed inefficient; to establish a regulatory agency in order to promote competition; and to privatise public utilities. Over the last two decades, Brazil and India underwent changes in their regulation inspired by these ideas. In this chapter, we examine the nature of these changes and their consequences for investment. After a brief presentation of the specific conditions surrounding electricity generation and distribution in both countries, we will examine their respective regulatory policies. Finally, we will try to evaluate their impact on investment.

The nature of electricity development in Brazil and India

Brazil and India share one important characteristic influencing the development of their electrical industry. Not only do they figure among the five most populated countries in the world, but their population is still growing, and this growth has actually accelerated over the last three decades. This goes hand in hand with more urbanisation. In 1995, 80% of Brazilians lived in cities; in India, only around one quarter of the population was urbanised by 1990, but this country must deal with gigantic urban centres, including Mumbai, Delhi and Calcutta.[1] This means that a continuous pressure of electricity demand over supply will prevail. However, both countries will not address this common issue in the same way.

Brazil has a more integrated energy transportation network than India. From 1970 to 1990, the installed capacity of its electric utilities grew from 10.3 to 52.1 GW and, in

1990, hydroelectricity accounted for almost 86% of this capacity. The Brazilian government encouraged the hydraulic option during the 1960s and 1970s.[2] The fact is that the country has an immense hydraulic potential. In 2006, its potential was estimated at 260 GW. The problem, however, is that hydroelectric generating plants are located far away from the consuming centres making an extended transportation system necessary: in 1995, Brazil had a 153,406-km high voltage transportation network.[3] Until the early 1990s, the development of electricity production has thus been strong. However, energy consumption is increasing at an even faster pace. From 1980 to 1990, final electricity consumption increased from 121.67 to 212.13 TWh. Brazil, which exported electricity in 1980, became a net importer a decade later and this new tendency will be accentuated in the following decades. Electricity consumption is exercising a continuous pressure over the generation of electricity. This situation is aggravated by the high amount of power loss in the transportation system: in 1990, losses are estimated at 31.7 TWh, representing 12.6% of the national supply.[4] This percentage is higher than normal.[5] Brazil suffers from energy theft especially in the favelas of big cities like Rio de Janeiro and Sao Paulo. The country consequently needs to invest considerably more in its electricity industry.

India's production falls very short of demand, too, in spite of the government´s significant efforts. During the 1980s, installed capacity more than doubled from 33.32 to 71.75 GW. Thermal electricity predominates in 1990, making up 72.3% of the total production (332 TWh). Coal remains the dominant source of electricity (65% of the total capacity in 1990). With 82.4 billion tons of proven reserve in 2001, India is the third largest producer of coal in the world. Gas as an energy source is on the rise, as is nuclear energy (5 stations in activity in 1992), albeit at a slow pace: these respectively represented 4% and 2% of the total capacity in 1990. Hydroelectricity takes second place, with 25.5% of total energy generation.[6] Its growth is much slower, since its plants are small compared to those in Brazil. It has been difficult to erect bigger ones for ecological as well as political reasons.

Since 1980, demand has systematically pressurised supply. Energy consumption increased by 70% between 1983 and 1999, while the population went up by only 29%.[7] In 1991, the Indian government made significant reforms to stimulate industrial production; this had a significant effect on energy consumption. Moreover, electrical appliances are beginning to replace servants in medium and high income dwellings, and so does the air conditioner over the ventilator.[8]

And yet, the per capita consumption remains very low, with 340 KWh in 1992. Many shortages or voltage failures occur, not to mention electricity theft at a widespread level: in 1990, electricity losses count for 17% of national production. Ten years later, only half of the electricity is billed and even less was paid for.[9] Needless to say, the Indian electricity industry must undergo major restructuring, and the changes of 1991 are favourable to setting the stage for such a change.

Despite the strong growth of their electrical industry, both Brazil and India need more energy, due to the rapid increase of consumers. To achieve that, massive investment is necessary. Many local analysts and journalists are becoming impatient about what is considered an unacceptably slow development of the electrical infrastructure. And from the developed countries came a new wind of reform, consisting of privatising and deregulating the public utilities. It was only a matter of time for these ideas to spread over the developing countries.

The reforms of the 1990s

Brazil and India implemented reforms during the 1990s so that investment – especially Foreign Direct Investment (FDI) – would be attracted to their electrical industry. Their pre-existing regulatory framework was proving deficient from the 1980s. However, the two countries took different paths to reform: Brazil underwent major reforms in the 1990s, whilst India introduced only gradual changes to its 1948 Electricity Act.

The reforms of 1995 in Brazil

The Brazilian economy was being seriously shaken during the 1980s and the 1990s. Low output per worker, a gigantic public debt and hyperinflation paralysed its economy.[10] Brazil had just come out of two decades under a military regime combined with a strong protectionist policy. The restoration of democracy and the opening of the economy to international competition rapidly exposed the obsolescence of its technology and of its working methods. Time had come for a renewal and international investment, so the successive governments made a point of adopting policies favourable to a broad opening of the economy. However, this process exposed Brazil to inflation; in response, the government froze the public utilities' tariffs in an attempt to control it and, consequently, seriously slowed down their chances of expansion. Electricity would suffer from this new reality. The annual investment in the power sector was on average 12.6 bn US$ during the 1980s, though this volume was lower after 1986. In 1990, this fell to 8.7 bn and five years later to only 4.3 bn US$. The government's part in the financing of electricity was around 20% in the 1970s, thanks to a federal tax on energy consumption. But during the next decade, its participation falls to under 5%. The companies, unable to increase their tariffs, become heavily indebted (a debt equal to 98.4% of their resources in 1989).[11]

Immediately after his inauguration as president in 1994, Fernando Henrique Cardoso prioritised the economic liberalisation and the fight against inflation. He introduced a new currency, the real, hoping this would put an end to the three-digit (sometimes four-digit) inflation; and he implemented a programme of privatisation aimed at reducing the public debt and stimulating free enterprise.[12]

Prior to 1995, electricity was predominantly managed by a single public company created in 1962, Centrais Elétricas Brasileiras S.A., better known as Eletrobras, a holding responsible for the technical and financial organisation of the bulk of production (through four inter-regional companies), and all transmission, and part of the distribution. This is considered the most important holding in Latin America.[13] In 1995, Eletrobras controlled 59% of the national installed capacity, while the companies owned by regional states (such as Companhia Energetica de Sao Paulo) held 36% and the private sector the remaining 5%. Local states and private companies hold a bigger share in the transmission and distribution of electricity.[14] Over them, there is a regulating agency, the Departamento Nacional de Aguas e Energia Elétrica (DNAEE), under the responsibility of the Ministry of Mines and Energy. Its authority over Eletrobras is quite limited, however.[15]

In 1993, the government allowed the companies to adjust their tariff according to their needs while advancing them 26 bn US$.[16] But Cardoso went further. In 1994–96, he gave the Conselho Nacional de Desestatização (National Council of Destatisation), under the supervision of the Banco Nacional do Desenvolvimento[17] (BNDES), the mission to privatise the electrical industry. There were many reasons behind this decision: to reduce the public debt; to open the Brazilian network to a continental interconnection; to stimulate competition; and to encourage further growth in the electrical sector.[18] A long series of regulations and laws were adopted between 1995 and 1998, with the aims of reinforcing the concession rules, separating generation, transmission and distribution from each other, allowing big consumers to choose their furnishers, putting the subsidiaries of Eletrobras in the national programme of denationalisation, creating an independent regulatory agency (Agencia Nacional de Energia Elétrica, or ANEEL, responsible, among other things, for tariff setting) in replacement of the DNAEE, and an independent system operator. Moreover, in order to prevent sharp increase of tariffs (and inflation), the long-term contracts are replaced by 'initial contracts' meant to last up to 2002. From then, it would be possible to allow tariffs to be influenced by the free market.[19] In other words, these reforms aimed at introducing private ownership and competition to the electricity industry, all under the supervision of an autonomous regulatory agency. Brazil was inspired by the British experience, known as the 'English Pool'. The companies, once privatised, were to receive a 30-year concession and compete with each other in the selling and buying of energy. Various rules discouraged concentration, be it vertical or horizontal, beyond 20% of the national electrical industry. It should be noted here that the interest of the consumer was put in the background, whilst the state of public finances was foregrounded.[20]

The government put BNDES in charge of the privatisation. The bank first privatised the distribution sector, with the intent to do the same with production and with distribution later on.[21] To avoid speculation, the bank did not list the companies on the stock market but decided to auction them instead. From 1995 to 2000, 21 distribution and 4 generation companies were privatised. The auction gave some 22 bn US$ to the government, while 5.1 bn US$ worth of debts were transferred to the private sector.[22] The privatised distribution companies covered more than 60% of the Brazilian market. Most of them were bought via joint venture. For instance, Light Servicos de Eletricidade S.A., which distributes electricity mainly in Rio de Janeiro, was privatised in January 1996. Its new owners were Électricité de France (11.35%), the AES corporation (11.35%), Houston Industries Energy (11.35%) and Companhia Siderurgica Nacional (7.25%).[23] Multinational firms such as AES and Tractebel, but also state-owned enterprises (SOEs) such as Électricité de France and Energias de Portugal made important investments in various Brazilian companies. Locally, groups like VBC Energy and pension funds trusts like Previ appear among the new investors.[24] It should be noted that Eletrobras remains a state-controlled company. In 1998, the government was enthusiastic about the results and plans to privatise generating companies. In 1999, ANEEL wished to concede plants totalizing 3,765 MW. However, it only conceded 200 MW.[25]

After 1999, private investment slowed down. According to UN-Energy data, a worldwide decline of investment occurred in the power sector in developing countries, falling from 43 bn $ in 1997 to 14 bn in 1999 and then again to 6.7 bn in 2002.[26] Moreover, Brazil was seriously hit by the Asian crisis. Despite privatisation, the country's public and commercial debts were still high. Brazil was forced to borrow 40bn $ from the IMF and, in exchange, reduced its public expenses. The country devaluated its currency in 1999 and again in 2002. This situation kept FDI away.[27] As this wasn't enough, Brazil had to reduce its electricity consumption by 20% in 2001. The various efforts to reduce inflation during the 1990s cancelled many electricity projects and forced the existing hydraulic reserves to be diminished increasingly in order to satisfy a growing demand: in 1994, the reservoirs were full by 92%; in 2001, they were down to 30%. Generating capacity could no longer meet consumption. By investing in the distribution sector, private investors neglected the generation sector. The only generation expansions which occurred after 1995 were those initiated before privatisation.[28] Many blamed the dependency on hydro-electricity. Already the government tried to encourage thermal production of electricity, but this failed to materialize.[29] The already integrated transmission system favoured hydroelectricity and this heritage from the past could not be ignored, not to mention that the cost for gas transportation had increased. Gas stations were producing electricity at 40$ per MWh, while hydro-electricity generated this at only $30.[30]

This crisis had immediate consequences. First, it put an end to privatisation. And second, the population replaced Cardoso by Luis Inácio Lula da Silva, the leader of the Partido dos Trabalhadores, as president of the country the following year.

Reform in India, 1991–98

In the early 1990s, the Indian regulatory framework was virtually identical to that set up by the passing of the Electricity Supply Act in 1948 (and its amendment in 1956). This Act created two institutions: a Central Electricity Authority (CEA), of national scope, and State Electricity Boards (SEB), which are regional organisations. The CEA's mission consists of elaborating a national policy for electricity, coordinating and, if necessary, acting as arbitrator of the activities of the SEBs and, finally, providing information relating to the generation, distribution and consumption of electricity. On their part, the SEBs, under the close supervision of State governments, organize the generation and distribution of electricity in their respective territories: they have to ensure there is enough power produced and carried to all the licensees; this means an SEB could launch or take control of any generating station deemed necessary to achieve its objectives. Moreover, the SEBs must encourage rural electrification. So, without necessarily resorting to nationalisation, the SEBs have a constricting authority over the licensees, both public and private. In 1990, of 25 states, 18 have a SEB. In 1998, the SEBs were responsible for 63% of electricity production in India, the remaining part being in the hands of other public and private companies. The ultimate goal of this organisation is to set up a real national grid interconnecting both thermal and hydro-electric power stations.[31] To sum up, the central and states governments have a controlling hand over the electrical industry in India.

However, no national grid emerged from this organisation. Local states maintain a strong hold over their SEBs. The country suffers from disequilibrium among its regions: Northern and Eastern states could not sell their surplus energy to others, thus wasting more than a million tons of coal in electricity generation. Moreover, the tariffs are over-subsidised, especially in the rural areas, while the SEBs made significant losses. The tariffs become a regular electoral issue; consequently, they remain low to the point they cannot cover adequately the production and distribution costs.[32] On the other hand, important public corporations were created between the 1960s and the 1980s in order to implement national policies and compensate for the SEBs' state of paralysis: for instance, Rural Electrification Corporation (1969), National Thermal Power

Corporation and National Hydroelectric Power Corporation (1975), Power Grid Corporation of India Ltd (1989).[33]

Nevertheless, in 1991, due to an external debt crisis, India underwent major economic reforms aimed at liberalising its market.[34] Electricity was clearly affected. Across the decade, many politicians, journalists and economists regularly denounced the SEBs and the subsidy system and demanded reform which would attract a bigger participation of private investment.[35] Among the first measures adopted in 1991, one encouraged private and foreign investment to participate in the electrical industry and another reinforced the regional load dispatch centres.[36] Meanwhile, after the refusal by the World Bank to support its power project in 1991, the state of Orissa divided its SEB into production, transmission and distribution companies in 1996 and sold the distribution ones to private interests.[37] The general idea was to encourage private investment to enter *en masse* to the electrical industry. This policy led to some ill-conceived projects, such as the Dabhol Power Corporation, a creation of Enron, which attempted to build an expensive liquefied natural gas plant in the state of Maharashtra. This became a political issue until the Maharashtran government reversed the agreement in 2001.[38] This experience saw FDI take a greater distance from the prospects of investing in the Indian electrical industry.

But the central government didn't give up on its desire to change the electrical setting. More important modifications were to come. In 1995, the government put forward a Minimal National Action Plan for Power which received its application through the Electricity Regulatory Act of 1998. This plan intended to coordinate the activities of the SEBs, set up a national energy policy and, more decisively, make electricity tariff a non-political decision. To do so, the government created two institutions independent from public interference: the State Electricity Regulatory Commissions (SERCs), responsible for retail tariff fixation, licensing and planning; and the Central Electricity Regulatory Commission (CERC), dealing with inter-state tariffs, licensing and planning but on the national level.[39] Their common mission was to render the electrical industry more efficient and to stimulate competition among companies interested in investing in a power project. In 1998, an amendment was added to the Electricity Law: it gave an independent status to transmission with the creation of a Central Transmission Utility (CTU) and States Transmission Utilities (STUs), both being public companies. Their mission was to coordinate and plan electricity dispatching within each state and at the national level, and attract private investment in transmission. In other words, the Indian government was planning the setting up of an authentic national grid.[40] The private sector reacted to this legislation by founding in 1999 the Power Trading Corporation, with equity holders such as Tata Power and the Industrial Development Bank of India (IDBI).[41]

Finally, in 2001, the National Council of Applied Economic Research, at the request of the Ministry of Power, wrote a draft of a Bill destined to gather all the numerous legislations into one law and, through the process, deepen the reform of the electricity sector. This Bill brought about a major transformation of electricity legislation from 2003.

To sum up, the Government of India shares with its Brazilian counterpart a desire to improve efficiency. However, the Indian approach is different from the Brazilian one. Privatisation is not at the centre of its concerns. From the Indian point of view, public ownership *per se* is not the problem. The real problem was the industry's submission to political considerations. The SEBs proved incapable of investing properly in transmission and distribution since the states stopped them from increasing the tariffs. Giving autonomy to fixing prices has become the *sine qua non* condition to set up a proper national grid. In Brazil, privatisation prevails because of the poor state of the public finances. However, both countries still had an enormous need for investment. And yet, an unfortunate project, Dabhol, helped dry up FDI towards this sector in India, whilst Brazil faced major crises in 1998 and 2002, helping to reduce FDI flows to its electrical sector. New solutions had to be found.

The reforms after 2000

India and Brazil changed their legislation relating to electricity in 2003 and 2004. The Indian government wanted to consolidate all those modifications introduced during the previous decade. The Brazilians were interested – on the contrary – to bring about further transformations, some contradicting those adopted during the 1990s.

The Indian Electricity Act of 2003

After three years of debate, an Electricity Act was finally voted in during May 2003. This was then completed by an Amendment later that year.[42] This Act kept many legislative changes adopted during the 1990s. Globally, it consisted of the following points:

1. Except for hydro and nuclear, the Act makes compulsory the de-licensing of power generation since most SEBs could not provide the necessary investment. It was important to free the new producers from the liabilities of the SEBs so that they could expand their installed capacity. At the same time, the Act de-licenses captive generation. Big consumers could generate their own electricity, sell their surplus and compete with the traditional power generators.

2. On the other hand, transmission and distribution remain licensed, except for governmental agencies and for rural electrification. In the latter case, the Act aimed at encouraging local cooperatives to be independent from the SEBs. Transmission would be supervised by the Central transmission utility (CTU) and by State transmission utilities (STU). Both were excluded from trading electricity and from power generation. They provide an open access to the transmission system they are responsible for. Attached to their functions are: the National Load Dispatch Centre (NLDC); the Regional LDC, dividing the country in 5 regions; and finally State LDC. For their part, the distribution networks are subjected to open access and allow competition as well as parallel distribution networking. Metering will become mandatory and energy theft will be fought and their perpetrators prosecuted. At the same time, an Accelerated Power Development and Reform Programme (APDRP) was established, with the mission of funding the improvement to be made on the distribution networks.

3. If we look at the regulation *per se*, three changes are worth mentioning. First, the Central Electricity Authority (CEA), created in 1948, lost some of its regulating capacity and became more of a technical adviser. It became autonomous whilst its members must remain neutral. Second, the CERC and SERCs act as independent regulatory commissions. They determine tariffs (although they should not interfere if tariffs are fixed by bidding), are responsible for the transmission, distribution and trading

licences. They have to make sure competition will not be affected by the dominant position of a company. There is a possibility for creating joint commissions for more than one state. The states can subsidise electricity produced or distributed on their territories, but in such a way that they do not question the autonomy of the regulating agencies and influence the tariff fixing. Finally, new institutions are created, such as the Appeal Tribunal of Electricity, a Central Advisory Committee, and a National Electricity Policy serving as a guide for future development and for the functioning of the power market.

In 2008, 14 states restructured their SEBs. Delhi and Orissa privatised their distributions networks. Eight states kept their SEBs in the transmission sector or as licensees. According to a 2006 study, these SEBs already have a good performance or have improved theirs. However, a debate prevails about the usefulness of unbundling the SEBs.

On the whole, the Indian Act of 2003 does not encourage privatisation as such. Its main goal is efficiency. If a SEB can reform itself, so much the better. However, be they private or public, the companies must not hinder competition, which is deemed as the main source of improvement. And finally, it is interesting to note how far the Act goes in correcting the negligence of SEBs in the past: it addresses issues such as the strengthening of distribution networks and the fight against electricity theft. Although Brazil has similar problems, its reforms prioritised other issues.

The Brazilian electricity law of 2004

In the aftermath of the 2001 crisis, the Cardoso government launched a programme to diversify electricity sources. Hydro-electricity was considered as one of the main agents of the system's collapse. Consequently, other sources of energy were favoured, such as gas-fired and nuclear plants. The programme, with the help of Petrobras, sought to build 55 natural gas-fired stations, representing 23,000 MW of installed capacity, and to continue with the construction of the nuclear station Angra III. Whilst the nuclear programme went ahead as scheduled, the gas-fired project failed. Hydro-electricity kept selling the cheapest electricity available, while gas importation was seriously hampered by a disagreement with Bolivia.[43]

In 2002, Lula da Silva replaced Cardoso at the head of the Brazilian government. With him came a new political vision with an explicit preoccupation for the poor and social justice. This meant, among other things, bringing electricity to the favelas and to more remote areas of the country. The government subsidised the electrification of favelas and stimulated the purchase of electrical appliances. These measures pushed up electricity consumption to such a level that it became vital to be connected to a reliable network and consequently to put an end to electricity theft. To achieve that, the government democratised consumption credit and kept prices low.[44]

This policy meant an increase in electricity generation and transmission.[45] Considering that investment was scarce, public regulation and planning became necessary. In March 2004, a new regulation was adopted by the Government, implying the creation of a 'Pool' (Ambiente de Contrataçao Regulado, ACR), coordinating demand and supply. The 'initial contracts', as designed by the reforms of the 1990s, would be replaced by long-term contracts. Electricity would be bought from the producers by the government and sold through auction to the distributors. In this way, hydro-electricity with its amortised costs would not be able to impose its prices on the new energy sector such as gas or nuclear. The risk is shared by the private participants and not only by the Government, who was more an auctioneer than a buyer: it imposed a tariff threshold and secured the supply of electricity. The pool should help the small customers (households and small and medium-sized enterprises (SMEs)), while large customers had the possibility to buy their energy directly from the producers in a free market parallel to the pool system. Moreover, large consumers can invest in production and sell the surplus; they have consequently a balancing role in the relation between demand and supply. The pool works in the following way: distribution companies must evaluate the amount of energy surplus they will require for the next 3 to 5 years and submit their needs to the governmental institution, the Empresa de Planejamento Energético (EPE), who will guarantee that supply meets their requirements. The pool will be the place where a uniform price will be fixed from the average of the prices negotiated during the auction. Distributors will be forbidden to buy electricity from producers who are vertically related to them: the pool system must prevail. To make sure they can correct inaccurate estimations, distributors also have the possibility to buy or sell electricity from each other. If they must pay more, they can increase their tariff if the difference is inferior to 5%; otherwise, they have to pay for the difference.

Brazil had to plan its electricity development over the long-term, meaning more hydro-electrical stations, as well as nuclear and gas production. In addition, transportation needed expansion and improvement. The Ministry of Mines and Energy surrounded itself with institutions to make sure the sector would function and expand correctly. EPE, as we have seen, is responsible for the quantity (and nature) of energy required to meet the demand and its growth. The Comitê de Monitoramento do Setor Elétrico (CMSE) was created to make sure there would be no shortages as in 2001; in other words, it guarantees energy security and that EPE's proposals can be met. The Camara de Comercializaço de Energia Elétrica (CCEE) controls the regulated market contracts, while ANEEL deals with the technical and financial aspects of the sales within transportation and generation. The Operator Nacionaldo Sistema (ONS) looks after the geographical interconnection and the expansion of transportation. And finally, the Conselho Nacional de Politica Energética (CNPE) coordinates the various ministries in order to implement policy relating to electricity. Inside this system, the role of the government is reinforced. Moreover, no further privatisation was planned for this sector.

The model favours producers over distributors since the prices are fixed in advance. More importantly, it stimulates the expansion of supply, which is precisely what an energy-hungry country like Brazil needs. It should be noted that countries such as Hungary, Thailand, Indonesia and Pakistan have already implemented such an approach.

The idea behind these changes is to secure electricity supply and further expansion. Brazil still has some 190 GW of unexploited hydroelectric capacity, according to estimations in 2008, and its transmission network needs development and repair. The intervention of the government is essential. We should add that the government did not give up its aim of diversifying energy sources. But this time, instead of promoting gas, it put forward renewable sources of energy, such as wind, with a special programme, Programa de Incentivo às Fontes Alternativas de Energia Eletrica (PROINFA).[46]

It is interesting to note that the reforms introduced after 2000 in India and in Brazil do not promote privatisation. The development of production and the improvement of transmission and especially of distribution have been prioritised. And public enterprises, so it seems, can perform as well as private companies. Have these new rules played on the investment in electricity as they were supposed to?

The nature of the changes and their impact on investment

The reforms introduced by India and Brazil were inspired by the major goals set out by electricity reformers in the developed countries of the world: these were driven overall by the aim of increasing efficiency by reducing the state's role in the industry. The two countries' electrical industries were both afflicted with financial losses. Moreover, their consumption was continuously growing and pressurising supply. It was crucial to increase the installed capacity and, consequently, more investment was required. The reform was precisely meant to achieve that.

However, beyond these similarities, country differences were important. Although both countries have a federal constitution, the central government is more powerful in Brazil than in India as far as electricity is concerned. Brazil developed hydraulic power and built an extended transmission network throughout its territory to carry the energy to consumers. India lacked a proper national grid, in spite of institutions which were supposed to install it. It became a priority for India to establish one which could save significant money to the industry. Besides, Brazil had a powerful public holding dominating the electrical industry, Eletrobras. India had a mixture of SEBs and public companies deprived of similar influence on the national industry.

Finally, the two countries shared a common weakness paralysing the expansion of their electrical industry: the difficulty of setting proper tariffs. The Brazilian government kept utilities tariffs very low as a way to fight inflation. Many Indian states forced their SEBs to maintain their prices below the cost of energy for political reasons. On one side, the central government policy was the problem; on the other, the local states politics were the problem.

The adoption of reforms in the electricity sector is to be understood with these different backgrounds in mind. Both countries' electrical sectors needed strong investment and more autonomy from public authorities. Brazil chose privatisation and a definite openness to FDI. The initial idea was to replace state intervention with competition and, eventually, to dismantle Eletrobras. India took a different path: its main goal was to de-politicise electricity and reinforce the national grid. Private investment was encouraged, but so was a better coordination of the existing institutions. Both countries established autonomous regulating agencies.

Implementing the changes was a hasty procedure leading to crises such as the 2001 energy shortage in Brazil and the Enron-Dabhol affair in India. These problems needed to be corrected. In Brazil, security of supply and a proper distribution of electricity to the poor incited the government to discontinue privatisation and to act as an auction agent between the producers and the distributors of electricity. In India, it was important to reinforce the autonomy of electricity from politicians, make sure competition would prevail and see to it that electricity theft, which was more widespread than in Brazil, would be adequately defeated. To achieve these goals, Brazil and India multiplied new public institutions, contradicting the initial stated aim of reducing the role of the state in electricity.

Table 1. Installed capacity, generation and consumption of electricity (1980–2006)

	1980	1990	2000	2006
Brazil				
Installed capacity (GW)	33.37	52.13	68.18	93.16
hydro (%)	82.42	85.94	86.53	76.06
nuclear (%)	0.00	1.26	0.96	2.15
renewable (%)	5.27	3.84	3.75	6.54
thermo (%)	12.31	8.96	8.75	15.25
Generation (TWh)	154.65	251.13	405.61	482.29
Imports - Exports (TWh)	-0.05	24.39	43.00	41.16
Losses (TWh and %)	16,46 (10,6%)	31,7 (12,62%)	63,57 (15,67%)	70,55 (14,63%)
Final consumption (TWh)	121.67	212.13	321.47	382.36
China				
Installed capacity (GW)	65.87	126.64	299.09	517.55
hydro (%)	30.85	27.31	24.40	22.68
nuclear (%)	0.00	0.00	0.72	1.32
renewable (%)	0.00	0.00	0.17	0.37
thermo (%)	69.15	72.69	74.70	75.62
Generation (TWh)	309.76	633.19	1,374.29	2,899.17
Imports - Exports (TWh)	0.31	1.84	-8.75	-6.88
Losses (TWh and %)	24,29 (7,84%)	42,84 (6,77%)	93,83 (6,83%)	181,67 (6,27%)
Final consumption (TWh)	261.49	549.34	1177.89	2528.95
India				
Installed capacity (GW)	33.32	71.75	108.15	143.77
hydro (%)	35.39	25.52	22.65	22.48
nuclear (%)	2.58	2.18	2.06	2.34
renewable (%)	0.00	0.03	1.00	4.31
thermo (%)	62.03	72.27	74.29	70.87
Generation (TWh)	140.59	332.01	684.20	892.23
Imports - Exports (TWh)	-0.04	0.91	1.35	2.81
Losses (TWh and %)	21,32 (15,17%)	56,52 (17,02%)	155,07 (22,67%)	188,92 (21,17%)
Final consumption (TWh)	97.90	219.88	375.39	517.21

Source: IEA

What was the impact of these new rules on the electrical industry? Table 1 shows this was ambivalent. In the Indian case, the installed capacity doubled from 1990 to 2006, but in the previous decade its growth was greater; in the Brazilian case, the results were even slower. If we compare these with the Chinese case, the Indian and Brazilian results remain modest. In 1980, the Chinese capacity was exactly double that of the Indian and the Brazilian capacities. In 2006, it was more than 5 times the Brazilian and 3.5 times the Indian ones. The sources of electricity have not significantly changed. Hydro-electricity has lost 10% in favour of other sources in Brazil, while in India the percentage remained more or less the same, except for renewables, which have gained 3%. In Brazil, nuclear and renewable account for 8.6% while thermal makes up 15.3%. There are two nuclear stations which have been added, while plants using gas and diesel have been established (especially in the isolated Northern areas). In India, wind energy was developed. More importantly, the growth of installed capacity has not been able to keep up with demand. In 2009, in Brazil, while the industrial consumption of electricity was reduced by 8% due to the international recession, residential and commercial consumption increased by 6% each.[47] The pressure on supply is still important. Brazil imports more energy than before, while shortages are still frequent. The energy losses are persistently high. *Light*, an important distributor in Rio de Janeiro, lost 30% of its energy through theft in 2007 alone.[48]

Needless to say, the two countries will require enormous investments in the years to come. According to *World Energy Outlook* (WEO), Brazil needs to invest 252 bn US$ in its electricity generation and distribution while India will have to pour 956 bn US$ into its own system from 2005 to 2030.[49] On average per year, this would mean 10 bn for Brazil and 38 bn for India. Is this a realistic assumption?

The national statistics are eloquent on that matter. If we look at figures 1 and 2, we have to conclude that the estimations are very ambitious. Figure 1 gives the annual investment in energy, both public and private from 1991 to 2010. Excepting the year 1997 and 2008–2009, these investments have not surpassed 10 bn $ per year. If we take into account that this figure includes all forms of energy, this means that the part of electricity is even smaller. Figure 2 shows annual FDI flowing into Brazil and India. One can see that FDI is very vulnerable to economic conjuncture, and only recently it went beyond 40 bn $ per year. Considering that India makes do with much lower investment than Brazil, the WEO's predictions look utopic. In the recent years, their economic performance and their reputation as members of BRIC certainly attracted a significant amount to be invested in their economy and energy. However, the global recession is dampening this trend.[50]

Did the new regulation help Brazilian and Indian electricity as far as investment is concerned? Figures 1 and 2 provide a complex answer. In the Indian case, regulation seems to have played a negligible role. The economic boom after 2005 had more influence on the investment on energy and on FDI. Initiatives from industrial groups played a decisive role in attracting international investment in electricity. For instance, in 2009, Tata Power was able to raise 335 million $ through global depositary receipt issues meant to be listed on the Luxembourg Stock Exchange.[51] The way investment comes into the Indian electricity seems however to depend more decisively on the capacity the nation has of mobilizing the various actors to make sure a national programme will work. In 2008–2009, India's target in increasing its electricity installed capacity was 7.53 GW. It reached only 46% of it. Delays in supplying equipment and also coal shortage were responsible for this situation.[52]

Figure 1: Investment in energy, 1991-2010

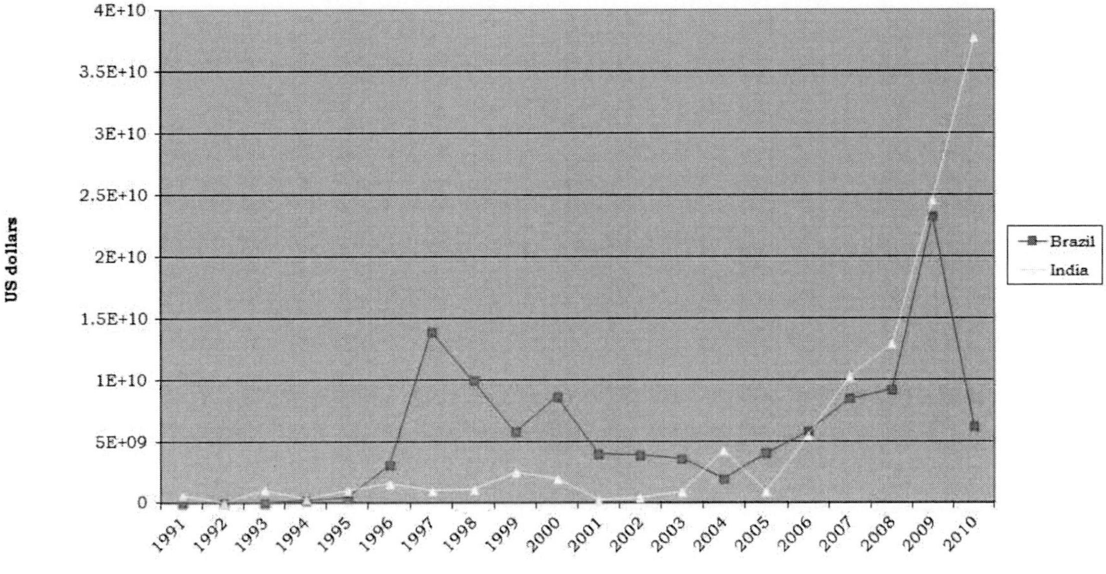

Source : World Bank

Figure 2: FDI in India and Brazil, 1991-2010

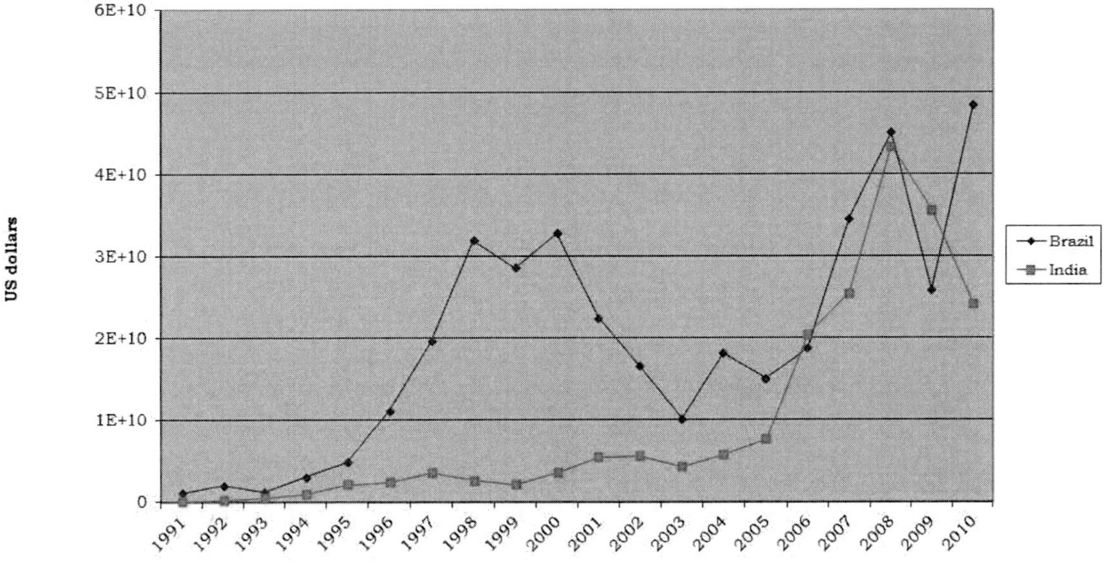

Source: World Bank

In the Brazilian case, on the other hand, the privatisations between 1996 and 2000 did play a positive role. From 1996 to 1998, FDI in the Brazilian electricity jumped from 9.9 to 22 bn \$.[53] In September 1998, the Belgian group Tractebel (which would become part of Suez group in 2003 and GDF Suez in 2008) acquired Gerasul, a generator of electricity, and renamed it Tractebel Energia in 2002. This group plays a significant role in the development of electricity in Brazil. In 2009, it inaugurated a 241 MW plant on the river Tocantins at São Salvador and is currently building two big plants at Estreito (1,087 MW) and Jirau (3,300 MW).[54] Eletrobras is controlled by the government up to 54% of its common shares (which form 80% of the total shares) and almost 20% of its preferred shares. The rest of the shares are in the hand of private investors. Among them, non-resident investors own 15.5% of the common shares and 47.7% of the preferred ones. Among the foreign shareholders, we have banks such as BNP Paribas and China Development Bank, with whom Eletrobras has signed a financial agreement worth 430 million US\$ for the building of phase C of the Candiota II thermal plant, owned by a subsidiary of Eletrobras.[55]

The 2004 Brazilian reform, however, failed to play such a positive role. In the short term, its implementation seemed to cool down investment. The attitude was a prudent one. But in the following year, investment rose, thanks to the stimulation brought by the upsurge of FDI.

Conclusions

Two main remarks should be stressed. First, analysts tend to study national regulation according to a predefined model set up by institutions, in particular, the World Bank. No doubt such a model has inspired many developing countries in the redefinition of their utilities' regulation. However, the reception of the model must also be taken into account. Each country has specific expectations from the new rules established in the developed world and these expectations will strongly orientate the way the model will be implemented. Brazil wanted to free its electricity from the government's fight against inflation: it privileged privatisation. Then, without questioning the results of privatisation, Brazil under Lula da Silva stopped the process and insisted on securing supply and providing energy for all. India wanted its electricity to evolve independently of local politics: it encouraged autonomous institutions, be they private or public, old or new, to take the lead in the sector.

Second, did the new regulation influence investment? In the case of Brazil, yes; in the case of India, no. Is Brazil closer to the international investors, and especially the Americans and Europeans? Possibly, although the examples of Enron and Tata show that India is considered a major place for investment. Privatisation may have played a significant role: Brazil benefited from a much higher FDI than did India. But the evolution after 2005 seems to demonstrate that this issue is not that relevant after all. Further study is needed to answer correctly that question, but it appears that the investment strategies of utility companies are of key importance here.

Notes

1 *World Energy Outlook, 1995 Edition*, p. 136; Census of India, Office of the Registrar General & Census Commissioner, Government of India, New Delhi.
2 Spiller (1996), p. 97; Oliveira (2007), p. 37.
3 Ferreira (1999), p. 8; Mendonça (1999), p. 75.
4 International Energy Agency (2002).
5 In Canada, a country with an elaborated transportation system, these losses are around 7–8%.
6 International Energy Agency 2002, pp. 31–38; Yadav, Surendra Singh and P.K. Jain, 'India's Power Sector: Need for a Better Management', in *Revue de l'Énergie*, no. 511, Nov. 1999, pp. 709–711; *World Energy Outlook, 1995 Edition*, p. 173.
7 Pachauri, Shonali, *An Energy Analysis of Household Consumption: Changing Patterns of Direct and Indirect Use in India*, Dordrecht, Springer, 2007.
8 *World Energy Outlook, 1995 Edition*, p. 192.
9 *World Energy Outlook, 2002 Edition*, p. 288; *World Energy Outlook, 2003 Edition*, p. 390; Tongia (2007), 117–127; Garg (2008), 70–71.
10 Adrogué (2006), p. 3.
11 Ferreira (1999), pp. 8–10; Calahorrano (2004), pp. 378–379; Adrogué (2006) pp. 7–9; *Brazil. How to Revitalize Infrastructure Investments in Brazil. Public Policies for Better Private Participation*, vol. 1, pp. 19–20 and vol. 2, p. 118; Spiller (1996), pp. 91–94; Tongia (2007), pp. 38–44; Arbache (2006), pp. 327–329; Vidal-Luna (2006), pp. 140–141; Baer (1998), pp. 511–512; Almeida (2009), p. 399; Mendonça (1999), p. 73.
12 Font (2003), pp. 21–23.
13 Defeuilley (1999), pp. 77–80; Tozzini (2006), p. 59; Baer (1998), pp. 508–511.
14 Ferreira (1999), p. 7; Mendonça (1999), pp. 75–76; Calahorrano (2004), p. 385.
15 Spiller (1996), p. 91; Ferreira (1999), p. 8.
16 Oliveira (2007), pp. 47–48; Ferreira (1999), p. 13; Baer (1998), pp. 513–514; Mendonça (1999), p. 74; Adrogué (2006), p. 4.
17 Known in English as Brazilian Development Bank.
18 Font (2003), p. 29; Mota (2003), p. 4; *World Energy Outlook, 1995 Edition*, pp. 121–124; Calahorrano (2004), p. 376; Oliveira (2007), p. 50; Mendonça (1999), pp. 76–77.
19 Law 8987/1995, Law 9074/1995, Decree 1503/1995, Law 9427/1996, Decree 2655/1998, Law 9648/1998.
20 Mota (2003), p. 35; Calahorrano (2004), p. 385.
21 Defeuilley (1999), Calahorrano (2004), pp. 380–385; Oliveira (2007), p. 50.
22 In all the economic sectors, from 1991 to 2001, Brazil privatised 119 companies; this gave the government 67.9 bn US$ and the transferring of the debt was evaluated at 18.1 bn. This represents the biggest privatisation in the world (Anuatti-Neto (2005), p. 145).
23 Moody's International Manual, 1998, vol. 1, p. 1053.
24 Ferreira (1999), pp. 27–28.
25 *World Energy Outlook, 2000 Edition*, p. 221.
26 UN-Energy (2005), p. 15.
27 Goldemberg (2004), pp. 22–23; Almeida (2009), p. 399; Sachs (1998), p. A23; Sader (2010), pp.50–52; Font (2003), pp. 33–35.
28 Calahorrano (2004), p. 385; Mota (2003), p. 8; International Energy Agency (2005), p. 25.
29 OECD (2005), p. 23.
30 *Ibid.*
31 *India, a Reference Manual* (1980), p. 280; *India Reference Manual* (1953), pp. 212–215; *Second Year of Freedom* (August 1948–August 1949), p. 92; *India 1990, A Reference Manual*, MIB, Government of India, New Delhi, p. 488; Garg (2008), pp. 2–4; Niranjan (2004), p. 4; *World Energy Outlook, 2000 Edition*, p. 312.
32 Jalan (1993), pp. 257–265; *World Energy Outlook, 1995 Edition*, pp. 200–201; Sachs (2001), pp. 56, 98–100; International Energy Agency (2002), pp.42 et seq.
33 Niranjan (2004), p. 5; Tongia (2007), p. 115.
34 Chandrasekhar (2006), pp. 250 and 256–259.

35 Yadav (1999), pp. 709–711; *World Energy Outlook, 2000 Edition*, pp. 307, 312–314; Sachs (2001), p. 290.
36 *World Energy Outlook, 1995 Edition*, p. 201.
37 Tongia (2007), p. 111; Sachs (2001), pp. 60, 100 and 283.
38 Tongia (2007), pp. 140–141; Beder (2003), pp. 302–312; Mehta (1999).
39 Tongia (2007), p. 111; Sachs (2001), p. 59; International Energy Agency (2002), pp. 50–53; Niranjan (2003), p. 4; Garg (2008), pp. 4–10; *World Energy Outlook, 2002 Edition*, p. 289.
40 International Energy Agency (2002), pp. 52–53; Sachs (2001), pp. 59–60.
41 Niranjan (2004), p. 5; International Energy Agency (2002), p. 54.
42 Niranjan (2004), all the book deals with the 2003 Act; Garg (2008), pp. 10–22; Tongia (2008), pp. 172–174; Wolak (2008), pp. 140–143. Two more amendments will be voted in 2007 and 2010: their main objective is to reinforce the legislation relating to electricity theft.
43 *World Energy Outlook, 2002 Edition*, pp. 314–315; *World Energy Outlook, 2006 Edition*, p. 480; Meritet (2004), pp. 181–182.
44 See Arbache (2006), pp. 340–347; *World Energy Outlook, 2006 Edition*, pp. 449–450; Smith (2009); Oyevinka (2010), pp. 68–70 and 100–101; Trudel (2010), pp. 34–35; Bartiromo (2009), pp. 13–15.
45 The following pages are based on: OECD (2005); Almeida (2009), pp. 398–404; Agência Nacional de Energia Elétrica (2010); *How to Revitalize Infrastructure Investments in Brazil,* vol. 2, pp.70–75; Meritet (2004), pp. 180–181.
46 Velcan Energy (2008); *World Energy Outlook, 2006 Edition*, p. 482.
47 EPE, Consumo nacional de energia elétrica na rede por classe, 1995–2011. Accessed 10 May 2012 at www.epe.gov.br/mercado/Documents/Consumo.
48 *How to Revitalize Infrastructure Investments in Brazil* (2007).
49 *World Energy Outlook, 2006 Edition*, pp. 482 and 486; *World Energy Outlook,2007 Edition*, p. 520.
50 Ragir (2011), pp. 42–43; Leahy (2011), p. 22; Sharma (2012), pp. 80–87.
51 'Tata Power raises $355m through GDRs', *The Economic Times Mumbay*, 23 July 2009, p. 8.
52 Nidhi Verma, 'Coal shortage, lack of new power capacity to cause 30% shortfall', *Mint*, 21 July 2009, p. 11.
53 *World Energy Outlook, 2000 Edition*, p. 210.
54 GDF Suez, Press release 4 February 2009, 'GDF Suez inaugure le barrage hydroélectrique de São Salvador au Brésil'.
55 Eletrobras, Annual Report, December 2007, pp. 37 and 44.

References

Adrogué, R., Cerisola, M., & Gelos, G. (2006). Brazil's Long-Term Growth Performance – Trying to Explain the Puzzle, IMF Working Paper, WP/06/282, December, 21 pp.

Agência Nacional de Energia Elétrica, *ANEEL 10 anos*, Brasilia, 2008, 129, last accessed 10 May 2012, at http://www.aneel.gov.br/arquivos/PDF/Aneel_10_Anos.pdf

Almeida, E. de, & Queiroz Pinto jr, H. (2009). L'évolution du cadre institutionnel des industries brésiliennes de l'énergie. *Revue de l'énergie*, 592, November–December, pp. 398–404.

Anuatti-Neto, F., Barossi-Filho, M., Gledson de Carvalho, A., & Macedo, R. (2005). Costs and Benefits of Privatization: Evidence from Brazil. In A. Chong and F. López-de-Silanes. *Privatization in Latin America. Myths and Reality.* Washington: World Bank Publications, pp. 145–183.

Arbache, J. S. (2006), Has Macroeconomic Policy Been Pro-Poor in Brazil? In A. Cornia (ed), *Pro-Poor Macroeconomics.* London and New York: Palgrave, pp. 326–348.

Asafu-Adjaye, J. (2000). The relationship between energy consumption, energy prices and economic growth: time series evidence from Asian developing countries. *Energy Economics, 22* (2000), pp. 615–625.

Aziz, J. (2008). Real and Financial Sector Linkages in China and India, IMF Working Paper (WP/08/95), IMF, 36 pp. Last accessed 10 May 2012 at http://www.imf.org/external/pubs/ft/wp/2008/wp0895.pdf

Baer, W., & McDonald, C. (1998). A return to the Past? Brazil's Privatization of Public Utilities: The Case of the Electric Power Sector. *The Quarterly Review of Economics and Finance*, *38*(3), Fall, pp. 503–523.

Bartiromo, M. (2009). How Brazil's Lula sees the emerging world order, *Businessweek*, October 12, 2009, pp. 13–15.

Beckerman, P., & Das, U.S. (1998). Risk-Capital Inflows, Inflation, and Macroeconomic Policy in India, 1993–95, *The Quarterly Review of Economics and Finance*, *38*(3), Fall, pp. 359–383.

Beder, S. (2003). *Power Play. The Fight to Control the World's Electricity.* New York/London: The New Press, xiv + 400 pp.

BNDES, Ministry of Development, Industry and Foreign Trade (2002). Privatization in Brazil, 1990–1994, 1995–2002, September, 74 pp. Last accessed 10 May 2012 at http://www.bndes.gov.br/SiteBNDES/bndes/bndes_en/Institucional/Publications_and_Presentations/studies.html

Brésil – Le rationnement de l'électricité électrocute la popularité du président Cardoso, last accessed 10 May 2012 at http://www.letinreporters.com/bresileco300501.html

Calahorrano, M. (2004). La déréglementation de la fin du 20e siècle en Amérique latine: le retour du secteur électrique au privé, *Revue de l'énergie*, *558*, July–August, pp. 376–388.

Census of India, Office of the Registrar General & Census Commissioner, Government of India, New Delhi.

Chandrasekhar, C.P., & Ghosh, J. (2006). Macroeconomic Policy, Inequality and Poverty Reduction in Fast-Growing India and China. In A. Cornia (ed), *Pro-Poor Macroeconomics.* London/New York: Palgrave, 2006, pp. 248–281.

Consumo nacional de energia elétrica por classe – 1995–2009. Empresa de Pesquisa Energética, Last accessed 10 May 2012 at http://www.epe.gov.br/mercado/Paginas/Consumonacionaldeenergiael%C3%A9tricaporclasse%E2%80%931995-2009.aspx

Consumo nacional de energia elétrica por classe – 1995–2009. Last accessed 10 May 2012 at http://www.epe.gov.br/mercado/Documents/Consumo%20nacional%20de%20energia%20el%C3%A9trica%20por%20classe%20-%201995-2009.pdf

Defeuilley, C., Cauret, L. & De Gouvello, C. (1999). Les concessions de distribution d'électricité au Brésil: régulation et recomposition industrielle. In *Annales des Mines*, August, pp. 77–80. Last accessed 10 May 2012 at http://www.annales.org/ri/1999/ri08-99/defeuilley77-80.pdf

Doire, M.-F. (2004). La coopération interaméricaine est-elle inappropriée face à la récurrence des crises énergétiques en Amérique latine? In Centre d'études interaméricaines (Université Laval), *Chroniques*, 13 June. Last accessed 10 May 2012 at http://www.cei.ulaval.ca/recherche_et_publications/chroniques/

Eletrobras (2007). Annual Report and Financial Statements 2007, December, 233 pp.

Eletrobras, Investment Projects in Electric Power Generation of PAC – In Progress Last Update: January/2010, last accessed 10 May 2012 at http://www.eletrobras.com/elb/data/Pages/LUMIS293E16C4ENIE.htm

EPE, Consumo nacional de energia elétrica na rede por classe, 1995–2011. Last accessed 10 May 2012 at www.epe.gov.br/mercado/Documents/Consumo

Ferreira, C. (1999). Privatizing the Electric Power Sector in Brazil. In BNDES, *Privatization in Brazil: The Case of Public Utilities*, Essay 6, April, 36 pp. Last accessed 10 May 2012 at http://www.bndes.gov.br/SiteBNDES/bndes/bndes_en/Institucional/Publications_and_Presentations/studies.html

Font, M. A. (2003). *Transforming Brazil: A Reform Era in Perspective.* Lanham: Rowman & Littlefield Publishers Inc., xiv + 272 pp.

Garg, A., Gaba, V., & Bajaj, J. L. (2008). *Regulation in Practice. Impact of Tariff Orders on the Indian Electricity Sector*, New Delhi: TERI Press, xvi + 240 pp.

GDF Suez, Press release 4 February 2009: GDF Suez inaugure le barrage hydroélectrique de São Salvador au Brésil. Last accessed 10 May 2012 at http://www.gdfsuez.com/fr/actualites/communiques-de-presse/communiques-de-presse/?limit=70&-annee_fin=2009&mois_fin=02&annee_debut=2009&mois_debut=02&thematique_a-vance=137&annee_avance=2009

Getting it together at last. A special report on business and finance in Brazil, *The Economist*, November 14, 2009, 18 pp.

Goldemberg, J., Lèbre La Rovere, E., Teixeira Coelho, S. (2004). Contributors: Osvaldo Soliano Pereira, Maria Silvia Muylaert, André Felipe Simões, Roberto Zilles, Patricia Guardabassi, Oswaldo Lucon, Expanding Access to Electricity in Brazil, 36 pp. Last accessed 10 May 2012 at http://www.afrepren.org/project/gnesd/esdsi/brazil.pdf

Goodman, J., & Dantas, I. (2010). The Lula Lieutenant Calming Business in Brazil, *Bloomberg Businessweek*, September 13–19, pp. 9–10.

India Reference Manual (1953), pp. 212–215; *Second Year of Freedom* (August 1948–August 1949).

India, a Reference Manual (1980). New Delhi: MIB, Government of India.

India 1990, A Reference Manual, New Delhi: MIB, Government of India.

International Energy Agency (2002). *Electricity in India. Providing Power for the Millions*, Paris: OECD/IEA, 129 pp. Last accessed 10 May 2012 at http://www.iea.org/

International Energy Agency (2005). *Saving Electricity in a Hurry. Dealing With Temporary Shortfalls in Electricity Supplies*, Paris: OECD/IEA, 130 pp. Last accessed 10 May 2012 at http://www.iea.org/.

Jalan, B. (ed.) (1993). *The Indian Economy. Problems and Prospects*, New Delhi: Penguin Books, xxiv + 368 pp.

Lane, P. R., & Schmuker, S. L. (2007). International Financial Integration of China and India. In L. A. Winters and S. Yusuf (eds), *Dancing with Giants. China, India, and the Global Economy.* Washington: A co-publication of the World Bank and the Institute of Policy Studies, pp. 101–132.

Leahy, J., & Pearson, S. (2011). Brazilian Assets Dumped despite Robust Economy, *Financial Times*, 10 August, p. 22.

Leloup, Y. (1960). La production d'énergie électrique au Brésil. *Annales de Géographie*, *69*(376), pp. 594–611.

Magnusson, L., & Ottoson, J. (2000). State Intervention and the Role of History – state and private actors in Swedish network industries.*Review of Political Economy*, *12*(2), pp. 191–205.

Mehta, A. (1999). *Power Play. A Study of the Enron Project*, Hyderabad: Orient Longman, xvi + 226 pp.

Mendonça A. F., & Dahl, C. (1999). The Brazilian Electrical System Reform. *Energy Policy*, *27*, pp. 73–99.

Meritet, S. (2005). Light, compagnie de distribution d'électricité au Brésil. *Revue de l'Énergie*, *566*, July–August, pp. 238–248.

Meritet, S. (2004). Vers une nouvelle crise de l'industrie électrique au Brésil? *Revue de l'Énergie*, *555*, March–April, pp. 179–182.

Moody's International Manual (1998), vol. 1.

Mota, R. L. (2003). *The Restructuring and Privatisation of Electricity Distribution and Supply Business in Brazil: A Social Cost-Benefit Analysis*, The Cambridge-MIT Institute, Massachusetts Institute of Technology Center for Energy and Environmental Policy Research, CMI Working Paper 16, DAE Working Paper WP 0309, 49 pp. Last accessed 10 May 2012 at http://www.econ.cam.ac.uk/electricity/publications/wp/ep16.pdf

Newbery, D. M. (2000). *Privatization, Restructuring, and Regulation of Network Utilities*, Cambridge, MA: The MIT Press, xviii + 467 pp.

Niranjan, R. S. (2004). *Guide to Electricity Laws in India. Based on The Electricity Act 2003*, New Delhi: Universal Law Publishing Co. Pvt. Ltd, xxvi + 358 pp.

OECD (2005). OECD Economic Survey of Brazil 2005: Regulation of the Electricity Sector. Paris: OECD.

Oliveira, A. de, (2007). Political economy of the Brazilian power industry reform. In D. Victor and T. C. Heller (eds), *The Political Economy of Power Sector Reform. The Experience of Five Major Developing Countries.* Cambridge: Cambridge University Press, pp. 31–75.

Oyeyinka, O., & Moreno, E. (dir.) (2010). *Sao Paulo. A Tale of Two Cities*, Nairobi, United Nations Human Settlements Programme (UN Habitat) / SEADE, *Cities and Citizens Series, Bridging the Human Divide*, no. 1, xxii + 157 pp. Last accessed 10 May 2012 at www.unhabitat.org

Pachauri, S. (2007). An Energy Analysis of Household Consumption: Changing Patterns of Direct and Indirect Use in India. Dordrecht: Springer, 211 pp.

Ragir, A., & Kopecki, D. (2011). Brazil's Credit Boom Could End in Tears, *Bloomsberg Businessweek*, 10–16 January, pp. 42–43.

Rodrigues, D. A. (2000). Investment in Brazil in the 1990s: Sectoral and Regional Views. In BNDES, June, 34 pp. Last accessed 10 May 2012 at http://www.bndes.gov.br/SiteBNDES/bndes/bndes_en/Institucional/Publications_and_Presentations/studies.html

Rothermund, D. (2008). *India. The Rise of an Asian Giant.* New Haven/London: Yale University Press, xiv + 274 pp.

Sachs, J., & Radelet, S. (1998). Next Stop: Brazil, *New York Times*, 14 October, p. A23.

Sachs, J. D., Varshney, A., & Bajpai, N. (eds) (2001). *India in the Era of Economic Reforms.* New Delhi: Oxford University Press, xxiv + 312 pp.

Sader, E. (2010). Huit années qui ont laminé le pays, in Là où le Brésil va…, in *Le Monde diplomatique* (coll. *Manière de voir*, no. 113), October–November, pp. 50–52.

Services urbains et développement durable, *L'analyse de six expériences de gestion déléguée dans le monde*, rapport de synthèse rédigé par Christophe Defeuilly et Dominique Lorrain, Ministère de l'Équipement, des Transports et du Logement, Direction des Affaires Économiques Internationales, IGD, ISTED, Paris, décembre 1999, 30 pp. to be found in http://www.isted.com/pole-ville/services_urbains/developpement-durable.pdf

Shalizi, Z. (2007). Energy and Emissions. Local and Global Effects of the Giants' Rise. In L. A. Winters and S. Yusuf (eds), *Dancing with Giants. China, India, and the Global Economy.* Washington: A co-publication of the World Bank and the Institute of Policy Studies, pp. 133–150.

Sharma, R. (2012). Bearish on Brazil. *Foreign Affairs*, May–June, pp. 80–87.

Smith, G. (2009). Brazil's Coming Rebound, *Businessweek*, August 17, pp. 42–45.

Spiller, P. T., & Martorell, L. V. (1996). How should it be done? Electricity regulation in Argentina, Brazil, Uruguay, and Chile. In R. J. Gilbert & E. P. Kahn (eds), *International Comparisons of Electricity Regulation.* Cambridge: Cambridge University Press, pp. 82–125.

Suman, B. & Singh, K. (2006). Domestic Financial Liberalization and International Financial Integration: An Indian Perspective. In J. Aziz, S. Dunaway, & E. Prasad (eds), *China and India. Learning from Each Other. Reforms and Policies for Sustained Growth.* Washington: IMF, pp. 145–199.

Tata Power raises $355m through GDRs, *The Economic Times Mumbai*, 23 July 2009, p. 8.

Taylor, R. P., Govindarajalu, C., Levin, J., Meyer, A. S., & Ward, W.A. (2008). *Financing Energy Efficiency. Lessons from Brazil, China, India, and Beyond.* Washington: IBRD/ World Bank, xviii + 285 pp.

Tongia, R. (2007). The political economy of Indian power sector reforms. In D. Victor & T. C. Heller (eds), *The Political Economy of Power Sector Reform. The Experience of Five Major Developing Countries.* Cambridge: Cambridge University Press, pp. 109–174.

Tozzini, S. (2006). As Conseqüências da Regulação Econômica sobre as Estratégias Empresariais das Concessionárias de Distribuição de Energia Elétrica – Brasil, Universidade de São Paulo, ProgramaInterunidadesdePós-Graduação em Energia, Sao Paulo, 263 pp.

Trudel, J. (2010). Un rêve nommé Brésil,*L'Actualité*, October 15, pp. 33–35.

UN-Energy (2005). *The Energy Challenge for Achieving the Millennium Development Goals*, 16 pp. to be found in http://www.un-energy.org/publications/50-the-energy-challenge-for-achieving-the-millennium-development-goals

Velcan Energy, Press release, 14 March 2008, Nouveau projet hydro-électrique au Brésil (16 MW) to be found in www.velcanenergy.com

Verma, N. (2009). Coal shortage, lack of new power capacity to cause 30% shortfall. *Mint*, 21 July, p. 11.

Vidal Luna, F., & Klein, H. S. (2006). *Brazil since 1980*. Cambridge: Cambridge University Press, xiv + 268 pp.

Wadja, R. (2001). Politics and power: Brazil's energy crunch, CNN, 11 August. Last accessed 2012 at http://archives.cnn.com/2001/WORLD/americas/08/08/energy.brazil/index.html

Wei, D., & Rose, A. (2009). Interregional Sharing of Energy Conservation Targets in China: Efficiency and Equity. *The Energy Journal*, *30*(4), pp. 81–111.

World Bank (2007). *How to Revitalize Infrastructure Investments in Brazil. Public Policies for Better Private Participation 2*, 2 vols, Document of the World Bank (Report no. 36624-br), 49 pp. (vol. 1) & 228 pp. (vol.2: *Background Report*). To be found in http://www-wds.worldbank.org/external/default/main?pagePK=64193027&piPK=64187937&theSitePK=523679&menuPK=64187510&searchMenuPK=64187283&theSitePK=523679&entityID=000020439_20070206110035&searchMenuPK=64187283&theSitePK=523679

World Bank (2010). Meeting India's demand for electricity. IBRD Results, March. Last accessed 10 May 2012 at http://siteresources.worldbank.org/news/resources/IndiaPowergrid4-1-10

World Energy Outlook, 1994 Edition (1994). Paris: International Energy Agency/OECD, 306 pp. Last accessed 10 May 2012 at http://www.iea.org/

World Energy Outlook, 1995 Edition (1995). Paris: International Energy Agency/OECD, xx + 380 pp. Last accessed 10 May 2012 at http://www.iea.org/

World Energy Outlook, 1998 Edition (1998). Paris: International Energy Agency/OECD, 478 pp. Last accessed 10 May 2012 at http://www.iea.org/

World Energy Outlook, 2000 Edition (2000). Paris: International Energy Agency/OECD, 458 pp. Last accessed 10 May 2012 at http://www.iea.org/

World Energy Outlook, 2002 Edition (2002). Paris: International Energy Agency/OECD, 532 pp. Last accessed 10 May 2012 at http://www.iea.org/

World Energy Outlook, 2003 Edition (2003). Paris: International Energy Agency/OECD, 515 pp. Last accessed 10 May 2012 at http://www.iea.org/

World Energy Outlook, 2006 Edition (2006). Paris: International Energy Agency/OECD, 600 pp. Last accessed 10 May 2012 at http://www.iea.org/

World Energy Outlook, 2007 Edition (2007). Paris: International Energy Agency/OECD, 663 pp. Last accessed 10 May 2012 at http://www.iea.org/

World Energy Outlook, 2008 Edition (2008). Paris: International Energy Agency/OECD, 575 pp. Last accessed 10 May 2012 at http://www.iea.org/

World Investment Report 2010. Country Fact Sheet: Brazil. Last accessed 10 May 2012 at http://www.unctad.org/Templates/Page.asp?intItemID=2441&lang=1

World Investment Report 2010. Country Fact Sheet: India. Last accessed 10 May 2012 at http://www.unctad.org/Templates/Page.asp?intItemID=2441&lang=1

Wolak, F. A. (2008). Reforming the Indian Electricity Supply Industry. In J. N. Bharwati & C. W. Calomiris (eds), *Sustaining India's Growth Miracle*. New York: Columbia University Press, pp. 115–165.

Yadav, S. S., & Jain, P. K. (1999). India's Power Sector: Need for a Better Management. *Revue de l'Énergie*, *511*, November, pp. 709–711.

From national monopoly to multinational corporation: How regulation shaped the road towards telecommunications internationalisation

Judith Clifton[a], Francisco Comín[b] and Daniel Díaz-Fuentes[a]

[a]*Facultad de Ciencias Económicas y Empresariales, Universidad de Cantabria, Santander, Spain;* [b]*Facultad de Ciencias Económicas y Empresariales, Universidad de Alcalá de Henares, Alcalá de Henares (Madrid), Spain*

One of the consequences of major regulatory reform of the telecommunications sector from the end of the 1970s – particularly, privatisation, liberalisation and deregulation – was the establishment of a new business environment which permitted former national telecommunications monopolies to expand abroad. From the 1990s, a number of these firms, particularly those based in Europe, joined the rankings of the world's leading multinational corporations. Their internationalisation was uneven, however: while some firms internationalised strongly, others ventured abroad much slower. This article explores how the regulatory framework within which telecommunications incumbents evolved over the long-term shaped their subsequent, uneven, paths to internationalisation. Two case studies representing 'maximum variation' are selected: Telefónica, whose early and unrelenting expansion transformed it into one of the world's most international of multinational corporations, and BT, whose overseas ventures failed and, with eroding domestic market share, forced the firm to partially retreat, becoming the least international of the large European incumbents. Long-term ownership, access to capital, management style and exposure to liberalisation strongly influenced firms' approaches to internationalisation.

Introduction

Across half a century, from the interwar period to the 1980s, the organisation of telecommunications systems across Europe reached considerable convergence (Millward, 2005, p. 245; Noam, 1992). Telecommunications had, by this time, been largely organised into national networks, and were operated by the state on a monopoly basis, often under the auspices of the Post, Telegraph and Telephone (PTT), usually as part of the Ministry of Post and Telecommunications (Bauer, 2010). Switching equipment was generally provided by private electrical equipment manufacturers, and terminal equipment by the PTT. The most conspicuous exception to this pattern of public involvement was Spain, where the Primo de Rivera dictator had allowed the US multinational corporation, ITT, to set up, control and operate a private monopoly from 1924 under the Spanish name Compañía Telefónica Nacional de España, known today as Telefónica (Comín & Díaz-Fuentes, 2004).[1] Other exceptions included Italy and Scandinavia, where the state only owned the trunk networks (Noam, 1992). Generally speaking, under state ownership and control, the management style of telecommunications monopolies was inward-looking, the priority being to provide national telecommunications services to users on a universal service basis (Clifton & Díaz-Fuentes, 2010a, 2010b). Millward (2005, 2010) provides a mix of political, legal, technological and economic reasons to explain the common economic organisation of telecommunications across Europe. If, originally, the potential for telecommunications was unclear, once their importance was fully recognised, governments sought to forge, and then, control, national networks for military and strategic objectives. Governments were also best positioned to resolve rights of way issues for network development. Monopoly operation of the national networks was facilitated by the relative simplicity of the technology – voice transmission to a single handset – and was, moreover, justified since the sector was understood to exhibit natural monopoly characteristics and network externalities. In the three decades previously, as telecommunications networks were evolving, multiple sets of conflicting interests had intervened in their development – geographically (local, regional and national) and through investment (private/public, domestic/foreign) – but these tensions were largely relaxed as the state assumed ownership and control.

Regulatory change, which, though somewhat unevenly and not always completely, undid this former arrangement, commencing from the end of the 1970s. Technological change such as the emergence of new customer premises equipment (computers, modems, multiple telephone handsets, fax and telex terminals) and value added services (mobile networks and phones, email, fax and audiovisual conferences) was a key driver of this shift. As telecommunications equipment became more important to users and, since it could lay no claim to natural monopoly status, pressure increased from the business sector, challenging the continued monopoly status of telecommunications and Public Telecom Operators (PTO) privileges (Millward, 2005, p. 252; Noam, 1992, p. 6). Major regulatory changes were introduced to reorganise the sector: privatisation, liberalisation and deregulation. Responsibility for implementing these changes was divided between the European, national and local authorities. In the context of the evolving European project, which aimed to forge a Single Market by 1992, the European Commission was responsible for implementing liberalisation and competition. Privatisation was a decision for national or local authorities, since the European Commission was bound to remain neutral on issues of ownership (Clifton, Comín, & Díaz-Fuentes, 2003). National governments were also responsible for the establishment of independent regulatory agencies to oversee telecommunications reform, since the formation of an European-level telecommunications regulatory agency was blocked by the 1958 Meroni doctrine, prohibiting as it did the upward delegation of decision-making of this type except that ratified by treaty amendment (Pelkmans, 2001, p. 455).

Now, the unravelling of the previous regulatory framework changed the business environment of national telecommunications monopolies, opening up new options for their managers. Under state ownership and control, expansion into foreign markets had been neither feasible nor necessarily even desirable (Clifton, Comín, & Díaz-Fuentes, 2007). The unfolding new regulatory framework opened the door to enterprise expansion abroad: indeed, many telecommunications enterprises seized the opportunity to internationalise swiftly and extensively. As a consequence, some of Europe's former national telecommunications monopolies emerged among the world's largest multinational corporations as measured by the United Nations Conference on Trade and Development (United Nations Conference on Trade and Development [UNCTAD], 2004, 2008). Resulting shifts in the economic organisation of the sector towards greater, albeit uneven, privatisation, competition and internationalisation, were partially reminiscent of the way in which telecommunications were organised in the first few decades from the end of the nineteenth century to around 1913, differing in that, in this new phase, the former telecommunications monopolies themselves became protagonists of change, leading the bids to compete in mergers and acquisition opportunities. This pattern mirrors closely the findings of Hausman, Hertner and Wilkins (2009) who contrast the role of electricity enterprises in the first and second wave of the sector's internationalisation.

But firms' paths from national monopolies to multinational corporations varied. While some incumbents, both large, such as Telefónica, France Télécom, Deutsche Telekom, and smaller, including TeleSonera, Telenor and Tele Danmark, emerged as successful, highly internationalised multinational corporations, others proved slower to go abroad, internationalising only modestly, whilst others met with failure, such as BT which, effectively, de-internationalised. Whilst scholars agree that regulatory change (privatisation, liberalisation and deregulation) of the telecommunications sector was a *prerequisite* for the ensuing expansion abroad of the former telecommunications monopolies (Bonardi, 2004; Sarkar, Cavusgil, & Aluakh, 1999), there is disagreement over why some firms internationalised much more vigorously than others. This topic is of considerable interest to scholars of business and policy history, and has important consequences for Europe's economic development. In this light, this article seeks to examine how regulation shaped, over the long-term, the paths taken by former telecommunications monopolies towards their transformation into huge multinational corporations. In the second section, we present the argument proposed by international business scholars which asserts that variations in the extent of internationalisation by former telecommunications monopolies can be explained by short-term asymmetries in national regulatory structures. Highlights of our results which examine data on the extent of firm internationalisation and indicators of regulation at country level over the short-term are presented, and our rejection of this argument is summarised. Instead, the extent of enterprise internationalisation is best explained at the firm level (Clifton, Díaz-Fuentes, & Revuelta, 2010). Over the long-term, however, regulatory structures influenced enterprises and their paths towards becoming multinational corporations. To illustrate, case studies are used. Following Flyvberg (2006), two case studies representing 'maximum variation' are selected: Telefónica, whose bold, early, unrelenting and deep internationalisation into Latin America and, later, Europe, transformed it into one of the most international of former telecommunications monopolies today; and BT, whose overseas' strategy, mainly comprising global alliances with North American and other peers met with failure, and its ultimate partial retreat, leading to a more modest presence abroad. We show in particular how enterprise ownership, rules on access to capital, management style and exposure to liberalisation all played a part in shaping firm behaviour, influencing its path to internationalisation.

Understanding how regulation shaped internationalisation

In the post-war decades, the bulk of the literature by international business scholars seeking to explain firm internationalisation built their theories largely around explaining those developments occurring to firms belonging to the manufacturing and oil sectors. This bias reflected the profile of multinational corporations, dominated as it was by these sectors (UNCTAD, multiple years). Of course, multinational corporations in telecommunications had existed for over a century: to cite just three giants, ITT, founded in 1920, became the main international telecommunications company of the United States; Cable & Wireless, created in 1934 on the merging of the Eastern and Associated Telegraphy Company & Wireless Telegraph Company, emerged as the international communications firm of the UK Post Office after its nationalisation in the post-war period; and Swedish Ericsson also embarked on internationalisation from the beginning of the twentieth century, offering telephone services in Latin American cities, such as in Mexico City from 1905 until, after fighting with its rival, ITT, saw all the network merge into Telmex in 1947 (Clifton, 2000). But these multinational corporations had not evolved from national monopolies; indeed, they often competed harder abroad than in their home markets. Broadly speaking, national telecommunications monopolies had little or no presence abroad during this period.

This situation changed dramatically from the 1990s, when a significant number of national incumbents emerged as huge multinational corporations as a consequence of regulatory reform from the 1970s. Parallel developments occurred in other sectors undergoing regulatory change – electricity, gas, water and transportation – though telecommunications was the most dramatic (UNCTAD, 2008). In response, international business scholars turned to re-develop existing internationalisation theories in the light of the particular economic characteristics of those utilities under analysis. In the case of telecommunications, scholars emphasised that the sector exhibited network economies and economies of scale, and also pointed out that investment opportunities in telecommunications tended to be large, rare or even once-off, high-risk and involved moving into spatially embedded networks. Concurring that regulatory reform, particularly privatisation and liberalisation, together with these economic features were central to understanding the internationalisation of the former national telecommunications monopolies, they integrated these policy and economic idiosyncrasies into their analyses (Sarkar et al., 1999, p. 368; Bonardi, 2004).

While scholars agreed that privatisation and liberalisation had been essential in establishing the conditions for the former telecommunications monopolies to expand abroad, explanation was needed on the dynamics and pace of this process. Why had some former monopolies embarked on a deep, unrelenting internationalisation drive, whilst others moved more slowly? Why were some internationalisation strategies successful, whilst other strategies failed repeatedly, resulting in retreating from their business abroad? Attention was placed on trying to explain the different *extent* of firm internationalisation. Among these studies, of particular importance are Bonardi (2004) and Sarkar et al. (1999) who proposed that telecommunications enterprises would internationalise at different paces depending on the differences in the timing and quality of privatisation and liberalisation policies across countries. Their explanation was as follows: because mergers and acquisitions in telecommunications are usually rare, risky, expensive and politically complex, those firms with significant political involvement and support have an advantage. Because governments tend to intervene more actively where levels of public ownership are higher, firms with more public ownership (or, less privatisation) would more easily access 'insider' information about the deals, and enjoy more support, knowledge and experience from politicians as 'diplomats' of the transactions. More privatisation, logically, meant the firm was external to 'insider' deals, perhaps making their expansion abroad more complex. A relationship between liberalisation and internationalisation was also established. Scholars including Bonardi (2004), Chari and Gupta (2008), Haar and Jones (2008) and Sarkar et al. (1999) argued that governments might opt (sometimes under pressure from the monopoly itself) to implement liberalisation slowly, or partially, to give breathing time to the 'national champion' to adjust to the changing circumstances, whilst providing it with monopoly rents in order to support their undertaking of risky operations abroad. Therefore, those telecommunications monopolies facing slower or less extensive pressures to liberalise at home would be better positioned to internationalise. Even in the European context, despite the fact that directives governed liberalisation, it was unlikely that all member states would implement liberalisation identically, regarding its timing, quality and extent. Some governments, including the UK, had introduced liberalisation before European directives were passed, whilst other governments (Greece, Ireland and Portugal) used formal extensions. To sum up, the argument claims that those former telecommunications monopolies least exposed to privatisation and liberalisation in their national settings would be those expected to internationalise most.

A straightforward means of testing these hypotheses is to assemble data on firm internationalisation and privatisation and liberalisation at the country level, and conduct regression analyses. This was performed elsewhere, and it was found that the extent of internationalisation is not determined by relative delays in privatisation or liberalisation and both hypotheses were thus rejected (Clifton et al., 2010). Significant investment abroad was undertaken both by firms that had experienced substantial inward liberalisation pressures, as well as those who were based in countries where liberalisation lagged behind. Similarly, firms with different levels of privatisation were able to go abroad successfully. Both smaller and larger telecommunications incumbents proved able to emerge as multinational corporations. The extent of internationalisation is, therefore, best explained at firm level.[2]

The diversity of the pace and timing of internationalisation activities of Europe's telecommunications incumbents is shown on Table 1. Of the large incumbents, by 2007, Telefónica was the most international, generating 64% of its total sales abroad, followed by Deutsche Telekom and France Télécom, at 53% and 50% respectively. BT was the least international of the large incumbents (15%). As regards timing, Telefónica was the first mover of the large incumbents, producing over 40% of its sales abroad by the end of the 1990s, followed by its German and French peers. BT was slower to go abroad, then, from 2002, it de-internationalised significantly, to return again, slowly, to overseas expansion. Being a smaller incumbent proved no barrier to strong internationalisation: by 2008, Telenor and TeliaSonera had proved particularly successful in their ventures, whilst others, such as OTE, Swisscom and TeleDanmark, internationalised more slowly.

Short-term differences in national regulatory frameworks do not explain these internationalisation patterns. The best indicators on the evolution of national regulation are constructed by the OECD (undated). Here, privatisation is simply measured as the proportion of incumbent shares owned by the government, whilst

Table 1. European telecommunications incumbents ranked by revenue: international sales as percentage of total.

Company	Country	1999	2001	2003	2005	2007	Revenues USD millions 2007–08
Deutsche Telekom	Germany	11.4	27.3	37.9	42.6	50.9	85,638
Telefónica	Spain	41.3	51.5	37.4	47.5	64.2	77,316
France Telecom	France	13.3	35.8	41.3	42.7	50.0	72,548
Telecom Italia	Italy	5.9	20.4	19.6	20.6	29.3	42,863
BT	United Kingdom	7.1	8.7	6.4	8.5	14.7	41,408
KPN Telecom	Netherlands	9.5	22.0	23.0	28.6	31.4	17,070
Telenor	Norway	15.4	10.0	36.8	51.3	65.5	15,780
TeliaSonera	Sweden	10.3	19.0	48.6	55.6	62.6	14,252
Swisscom	Switzerland	9.1	29.9	30.9	0.6	15.1	9241
OTE	Greece	0.5	1.9	19.3	24.5	25.5	8657
Portugal Telecom	Portugal	8.8	24.5	24.0	34.3	44.7	8422
TDC	Denmark	41.8	55.4	45.3	49.0	37.0	7228
Telekom Austria	Austria	0.0	7.1	11.0	25.5	35.9	6738

Source: Firm Annual Reports (1999–2008).

liberalisation is understood in two ways: first, it is measured by whether liberalisation legislation has been introduced (white for monopoly, grey for duopoly and dark grey for competition). Second, it is measured by the market share new entrants enjoy of the trunk telephony market ('real' competition). These liberalisation indicators are superimposed on Table 2. From the post-war to 1995, telecommunications was organised as a monopoly across Europe, with the exception of the UK, where a duopoly was established from 1984 to 1991, in which Mercury gained access to less than 20% of the market. Private sector involvement did not start with BT, as is commonly stated, but with Telefónica, which was established in 1924 as a private company and never fully nationalised, as is discussed later in the paper. By 1997, BT, Telefónica and TeleDanmark were fully privatised whilst average government ownership in the EU-15 was 66%. The UK was the liberalising pioneer, followed by Denmark, Finland and Sweden, from 1995, the Netherlands from 1996 and France from 1997, while nine other countries liberalised in time with the deadline set by the European Commission (1998–2001), including Spain, which had already introduced a weak duopoly from 1995. Despite this, Telefónica dominated its home market more securely than any of its peers, facing the lowest market penetration of new entrants in the trunk market.

How, then, can the different extents of incumbent internationalisation be accounted for?

Given our findings that internationalisation is best explained at firm level, one option is to conduct detailed case studies of individual incumbents and analyse how factors such as privatisation and liberalisation shaped the firm's approach to internationalisation. Following Flyvberg (2006), we design our case study using the 'maximum variation' strategy, whereby cases are selected that are the most similar regarding the main factors under examination bar that factor which needs explanation. It is therefore of interest to analyse Telefónica, whose early, aggressive and unrelenting expansion transformed it into one of the most international of incumbents today; and BT, whose hesitant and troubled ventures abroad led to its ultimate de-internationalisation, and a relatively modest international presence.

Table 2. Regulation of European telecommunications: liberalisation, market competition and privatisation, 1975–2007.

		1975	1980	1985	1990	1995	1997	1999	2001	2003	2005	2007
Austria	Market %	0	0	0	0	0	0	15	54	46	45	44
	Privat. %	100	100	100	100	100	100	88	61	47	30	27
Belgium	Market %	0	0	0	0	0	0	5	15	15	31	28
	Privat. %	100	100	100	100	51	51	54	54	54	54	54
Denmark	Market %	0	0	0	0	0	5	38	36	37	39	34
	Privat. %	100	100	100	100	51	0	0	0	0	0	0
Finland	Market %	0	0	0	0	60	60	62	63	63	62	60
	Privat. %	100	100	100	100	100	100	77	53	19	14	14
France	Market %	0	0	0	0	0	0	20	36	38	40	42
	Privat. %	100	100	100	100	100	62	62	59	59	43	27
Germany	Market %	0	0	0	0	0	0	28	33	40	43	49
	Privat. %	100	100	100	100	100	63	60	46	43	37	31
Greece	Market %	0	0	0	0	0	0	0	2	21	24	27
	Privat. %	100	100	100	100	96	65	58	42	34	31	28
Ireland	Market %	0	0	0	0	0	0	0	0	0	22	32
	Privat. %	100	100	100	100	100	80	41	0	0	0	0
Italy	Market %	0	0	0	0	0	0	15	25	25	27	29
	Privat. %	100	100	100	100	50	5	4	4	0	0	0
Luxembourg	Market %	0	0	0	0	0	0	2	7	12	18	24
	Privat. %	100	100	100	100	100	100	100	100	100	100	100
Netherlands	Market %	0	0	0	0	0	0	16	24	40	34	28
	Privat. %	100	100	100	100	56	44	44	39	19	8	0

(continued)

Table 2. (*Continued*).

		1975	1980	1985	1990	1995	1997	1999	2001	2003	2005	2007
Norway	Market %	0	0	0	0	0	0	12	27	32	31	30
	Privat. %	100	100	100	100	100	100	90	78	62	62	54
Portugal	Market %	0	0	0	0	0	0	0	9	9	20	31
	Privat. %	100	100	100	100	73	25	13	3	7	7	8
Spain	Market %	0	0	0	0	0	0	14	18	18	20	21
	Privat. %	35	35	35	35	21	0	0	0	0	0	0
Sweden	Market %	0	0	0	0	5	17	14	31	31	37	42
	Privat. %	100	100	100	100	100	100	85	71	46	48	50
Switzerland	Market %	0	0	0	0	0	0	19	37	45	42	38
	Privat. %	100	100	100	100	100	100	68	64	63	59	55
United Kingdom	Market %	0	0	0	8	19	24	36	46	46	43	40
	Privat. %	100	100	50	50	1	0	0	0	0	0	0

Liberalisation framework: Monopoly Duopoly Competition

Market %: Market share of new entrants in the trunk telephony market.
Privat. %: Percentage of incumbent shares owned by government.

Telefónica and BT represent large incumbents, both were exposed earlier on to privatisation than the other large incumbents, and both were subject to timely or premature liberalisation policies. This is done in the next section.

Longer-term regulation and incumbent internationalisation

This section aims to shed light on how longer-term regulation influenced Telefónica's and BT's approach to the opportunities opened to them by regulatory change, particularly, their different paths to becoming multinational corporations. In order to do so, the previously presented data on firm internationalisation, and country liberalisation and privatisation is supplemented with recently published important business history studies on both firms (Calvo, 2010; Parker, 2009), as well as comparative historical and economic studies of utilities and multinationals in Europe, company annual reports and other secondary material.

Telefónica

By the 1950s, Telefónica had emerged as Spain's largest firm but, by the 1990s, it had emerged as a leading world multinational corporation (UNCTAD, 2008). Historians have argued that Telefónica represents an exceptional case of business organisation in Europe (Calvo, 2010). Three major, interrelated, features help explain its distinctiveness over the long term, particularly vis-à-vis its European peers: firstly, its origins as a private, foreign owned and controlled firm; secondly, its management style, which closer resembled a private than a public enterprise; and thirdly, its relatively fluid access to credit and the stock market. Analysis of Telefónica is divided into three parts: origins, access to finance, and the road to internationalisation.

Born a foreign-owned and controlled private monopoly

The origins of today's Telefónica go back to 1924 when, following the *coup d'état* in 1923, General Primo de Rivera granted a 20-year contract to the recently founded Telefónica to operate the national telecommunications network and supply its equipment on a private monopoly basis. Telefónica had been founded just a few months before the coup and, despite having a Spanish name and its inclusion of well-known Spanish industrialists and bankers on its board, it was actually controlled by the emerging United States multinational corporation, ITT, working in conjunction with Spanish banks Urquijo and Hispano Americano (Álvaro, 2008; Comín & Gálvez, 2006, p. 119). ITT, in turn, had been established in 1920 by the Behn brothers who had merged two small telecommunications firms in Cuba and Puerto Rico. The ambitious brothers dreamed of achieving overseas what AT&T had accomplished in the US (Wilkins, 1974) and Telefónica became 'the jewel in ITT's communication empire' (Little, 1979, p. 450), facilitating ITT's entrance into Europe, and bolstering its stature as the world's leading telecommunications multinational corporation by the end of the 1920s (Sobel, 2000). Lacking technological capabilities of its potential competitors (AT&T owned Western Electric whilst Ericsson, Siemens-Halske and New Antwerp Telephone and Electrical Works dominated European markets), the Behns acquired International Western Electric company, the international subsidiary of AT&T and Western Electric, renaming it International Standard Electric Company, and Standard

Eléctrica, in Spain (Wilkins, 1974). By so doing, they diversified their business to manufacturing, in addition to sealing control over network and equipment provision to the Spanish market.[3] ITT was a key player in the 'Americanisation' of Europe (Zeitlin & Herrigel, 2004; Schröter, 2005), but Spain represented an extreme case, as most of ITT's investment in Europe was concentrated in the telecommunications manufacturing.

With the Civil War over, Telefónica's ownership changed. The Franco dictatorship immediately introduced tighter restrictions on FDI and multinationals, with the 1939 law specifying a 75% quota of Spanish national participation for companies (Wilkins, 1974). Many US multinationals, including ITT, found the political situation so complex and volatile that they sought to exit Spain (Álvaro, 2007, 2008). When Telefónica's contract expired in 1944, the government issued a decree placing the majority of stock in Spanish hands. So, despite having supported the fascists and the Nazi regime during the Civil War, the Behn brothers did not receive preferential treatment from the dictatorship (Little, 1979). Telefónica was nationalised in May 1945. The state took just over half of ITT's 80% stock; 12% went to the Central Bank, Banco de España, and the rest was split between 700,000 smaller shareholders. ITT retained the right to supply equipment on a monopoly basis, a privilege that would continue until the 1980s. ITT would, anyway, shift away from service and towards manufacturing provision (Sobel, 2000, pp. 7, 117).

Though Telefónica was nationalised whilst retaining monopoly privileges, as so many other telecommunications enterprises in Europe, the way in which it was nationalised differed to most of its peers. Nationalisation was never total; rather, Telefónica remained a mixed public-private ownership until its full privatisation in the 1990s. Comín and Gálvez (2006, p. 121) referred to this process as 'naturalisation', given the emphasis on indigenous ownership, whether private or public. This arrangement had important consequences for Telefónica's management style, which reflected conflicts of interest between private shareholders and public service investment. Indeed, Telefónica was managed more along the lines of a private business than a public utility. Unlike most of its European counterparts, it was never run as part of the Ministry or the Post Office. Instead, the regime colluded with private interests in the board (banks and ITT as manufacturer) at the expense of other shareholders. The regime appointed Telefónica's president, who decided on strategy, whilst the firm was organised according to clear functional divisions, with business managers reporting to the board and shareholders. Managers enjoyed high levels of autonomy.

Financing Telefónica

Thanks to its ownership characteristics, Telefónica became very successful at generating finance through public offerings. Between 1924 and 1966, 14 public offerings were made, attracting a total of 1.4 billion pesetas in shares (Calvo, 2010). Between 1950 and 1980, Telefónica leveraged external bank borrowing and self finance to launch new share issues. Its mixed ownership limited additional share emissions because the state was contractually obliged to purchase a percentage of issued shares (Calvo, 2010, pp. 261–262). From the mid-1960s, what became an infamous media campaign on Spanish television promoted popular capitalism through the sales of so-called Matildes, named after the girlfriend of the advert's protagonist, who tried to use his ownership of Telefónica shares as a means of demonstrating his prowess and credentials as a husband (Rodríguez, 2005). In reality, however, the rate of return and dividends were modest (6.74% in 1950 to 8.72% in 1977). The price of Telefónica shares remained stable until 1973; indeed, investing in Telefónica was considered so safe that shares were nicknamed 'widow's shares' in Spain, and categorised as 'blue chip' on the stock markets. However, after reaching a peak of 490 pesetas per share in 1973, their value plummeted to a low of 84.5 pesetas over the next four years, in parallel with the international economic crisis and the collapse of the Spanish stock market as the Franco regime came to an end.

The road to internationalisation

The democratic transition (1976–1982) marked a new phase in Telefónica's evolution, which was consolidated when Spain joined the European Community in 1986 (Comín, 2008). After accession, Spain would be forced to adopt Community rules on anti-trust and market competition. Even though European reforms would not affect telecommunications until the 1990s, and privatisation was not a competence of the European Commission, the Spanish government embarked on industrial restructuring and partial privatisation in the quest for improved performance. One main objective of the first Socialist governments (1982–1996) was to improve the management of monopolies, including Telefónica, partly, through its policy of partial privatisation. The aim was to retain public enterprises only in those sectors where their activity was considered strategic and there was some prospect of survival. The privatisation of public enterprises got under way from the mid-1980s when minority share packages of profitable public enterprises, including Telefónica, Repsol, Endesa and Argentaria, were placed in national and international stock markets (Comín, 2008).

As in the rest of Europe, the Spanish government recognised that significant investment was required to adapt to new communication technologies and to improve waiting lists for connections and service quality. At the end of 1982, Socialist MP Luis Solana was appointed president of Telefónica. Solana had long been an outspoken critic of Telefónica's management from the opposition, particularly its technological dependence on ITT (Solana, 1979). Solana broke the long-standing technological dependence on Standard Eléctrica, by introducing new policies on equipment acquisition and technology transfer. He publicly accused Standard Eléctrica of overpricing ('El negocio de ITT en España', 1983), whilst because ITT was diversifying away from telecommunications equipment provision to other businesses, the technology it supplied was in decline (López, 2003).[4] A breakthrough towards internationalisation and further privatisation occurred in 1985, when Telefónica became the first Spanish enterprise to be listed on the London Stock Exchange and other European markets. Telefónica embarked on various R&D and export-promotion projects, in addition to joint ventures and cooperative agreements involving industrial production and technology transfer. In 1986, strategic agreements and joint ventures were established with various international firms, including AT&T Technologies, SysScan, Olivetti, Brown Boveri, Philips, Saab-Scania, Telfin and Fujitsu. In June 1997, the government launched a public offering of 6% on the New York Stock Exchange, reducing state ownership from 38 to 32%. Its trajectory of attracting finance reduced the state burden to secure infrastructure investment finance. So, during the 1980s, whilst most European telecommunications enterprises – except for British Telecom – were still state-owned, Telefónica was subject to international external discipline through flotation in international stock markets. The aim behind privatisation was to have Telefónica listed in international stock markets, hence facilitating the finance of its international expansion without affecting the public budget (Bel & Trillas, 2005).

Solana was replaced by Cándido Velázquez, former president of Tabacalera, another historical monopoly, who embarked on a new phase of corporate internationalisation from 1989. Telefónica targeted acquiring telecommunications enterprises put up for sale in the aftermath of the debt crisis in Latin America, many of which became privatised monopolies (Ramamurti, 1996). Though Telefónica would later liberalise in line with European telecommunications directives, it is notable how its earlier acquisitions were achieved as a national monopoly. Telefónica acquired controlling stakes in Compañía Telefónica de Chile and Telefónica Argentina SA in 1990, and bought stakes in Compañía Anónima de Telefonos de Venesuela and Telefónica of Puerto Rico in 1991. In 1994, it bought participations in Colombian COCELCO and Telefónica de Perú and, in 1996, in Brasilian Companhia Riograndense de Telecomunicaçoes. To manage these International acquisitions, Telefónica International SA (TISA) was set up in 1994, becoming Telefónica's largest subsidiary. At the same time, Telefónica sought to develop international alliances to head off hostile takeovers and to gain momentum for expansion. Most of the strategic alliances in which Telefónica was involved were of a protective nature, between small and medium-sized enterprises. In 1993, Telefónica joined the UNISOURCE alliance along with KPN, Telia, Swiss Telecom, and AT&T Worldpartners. TISA remained outside of UNISOURCE since it was negotiating an agreement with Concert (BT 75%, MCI 25%). AT&T questioned conflicts of interest between Telefónica in UNISOURCE and TISA in Concert; in any case, neither alliance was to last.

With the accession of the Popular Party (PP) to power in March 1996, the drive to privatise intensified, becoming total and generalised in scope. For the first time in Spanish history, a government spoke openly about eliminating all public ownership (Comín, 2008). In June 1996, Prime Minister José María Aznar appointed José Villalonga as Telefónica's president, an old school friend of his, but who had no experience in the telecom industry (Bel & Trillas, 2005). Telefónica was fully privatised in February 1997, when the government sold its remaining 20% share (Clifton et al., 2007). Villalonga's team saw UNISOURCE as a failed international alliance so, in April 1997, Telefónica sought to join the Concert alliance, formed on the merger of BT and MCI in 1996 for two main reasons: to protect its home market from BT, which was at the time Telefónica's most serious rival in Europe, and also to avoid competing with MCI, one of its competitors in Latin America. When MCI was taken over by World Com in 1997, however, Concert collapsed, so Telefónica formed an alliance with MCI WorldCom to offer services in Europe and improve its position in Latin America. During Villalonga's short reign, until 2000, Telefónica continued its expansion across Latin America, increasing its stake in Telefónica de Argentina (28.8% to 97.9%), Telesp (17.5% to 86.6%), Telefónica de Perú (40% to 93.2%), and Tele Sudeste (17.6% to 75.6%).

From 2000 onwards, the PP apointed César Alierta as Telefónica's CEO. Alierta had experience in managing huge mergers, having been previously CEO of former Spanish tobacco monopoly, Tabacalera, when it merged with its French peer in 1999. The Socialist Party reappointed him in 2004, indicating Telefónica's growing autonomy from politics. During this period, Telefónica consolidated its international presence, staying in Latin America whilst also growing in Europe and has established joint ventures in China. It became the most global of all former telecommunications monopolies, and is ranked in fourth position world-wide in the sector by market capitalisation, number one among European integrated operators, and number two in the Eurostoxx in 2010. It is the leader provider in Brazil, Argentina, Chile and Peru and has substantial operations across most other Latin American countries, especially after acquiring BellSouth's assets in the region in 2004. In addition, Telefónica expanded in Europe, acquiring the incumbent Cesky Telecom in 2005 and acquiring a license for mobile telephony in Slovakia in 2006. It also bought O2, BT's subsidiaries operating in the United Kingdom, Ireland and Germany in 2006, whilst an alliance was signed with Telecom Italia (10.49% of voting rights) in 2007.

BT[5]

Origins as a state-controlled, government-run monopoly

From its origins in the late 1870s, telecommunications provision across Britain was mixed, being provided by private sector companies, such as the National Telephone Company, and the public sector, particularly, the General Post Office (GPO), which formed part of the civil service bureaucracy. In 1896, the GPO took over the trunk service and, by 1912, it had taken over the entire system, excepting a few local authorities, thus unifying the network for the first time and effectively installing a public telecommunications monopoly. Telecommunications in Britain henceforth evolved under state ownership and control, unravelling only from the end of the 1960s. The pace and level of development of telecommunications in Britain from the 1880s to 1950 followed the European average (Millward, 2009, p. 102). During this period, numerous debates ensued as to whether the Post Office should be converted into a nationalised industry rather than its organisation under the auspices of a government department. During the 1930s, the GPO was criticised as being too bureaucratic and 'Edwardian' in its running of the network and, though the Bridgeman Committee was set up in 1932 to investigate options for change, nothing materialised (Committee of Enquiry on the Post Office, 1932). Criticism sharpened during the 1950s as waiting lists for telephones rose to unacceptable levels but, again, no organisational changes were agreed. Lipartito (2004, p. 155) has argued that the fundamental problem plaguing the British network was long-term under-investment: the GPO was used by the government of the day as a tool of demand management, setting budget priorities according to macroeconomic needs until 1961. Furthermore, the failure to prioritise the national network was due to 'powerful cultural context' (Lipartito, 2004, p. 161). Run by civil servants, it was assumed telecommunications were something a person either wanted, or did not; feeble efforts were made to commercialise or expand them, in comparison with other leader nations. Britain had instead historically prioritised other forms of communication, particular, international cable and telegraph services, for both geopolitical reasons (controlling its Empire) and for related business reasons (to serve its foreign investments and multinational corporations). Regarding its running of the national network, the GPO behaved as a 'discriminating monopoly', under-investing in low-profit services and taking profits where they were made. London was the main priority: with the densest population and the most important businesses, scarce resources were concentrated on the network, supporting the city's development as a commercial, financial and residence capital (Lipartito, 2004, p. 157).

The road to privatisation

The Labour Party which came into power in 1964 pushed for reform, and finally opted to split Telecommunications from Post via the Post Office Act, 1969. At the same time, the Post Office ceased to be a government department, being established as a public corporation, a limited liability company whose shares were listed on the stock market. However, the Treasury still controlled the budget while the old functional management structure remained (Lipartito, 2004, p. 163). By the 1970s, the potential for technological change, such as digitalisation and convergence of voice, data and image, were starting to be recognised, whilst official calculations of the required investment to modernise the British network were of increasing concern to politicians, making external finance look increasingly essential (Parker, 2009, p. 241). The Carter Committee Report in 1977 recommended Telecommunications and Post be split further, into two individual corporations, leading to the establishment of British Telecom in 1980.

The real watershed in the evolution of telecommunications occurred in the 1980s. Initially, privatisation did not figure significantly in Conservative Party leader Margaret Thatcher's manifesto, the surprising success of selling council houses from 1979 prompted the government to embark properly on privatisation (Clifton et al., 2003). The Conservative Party calculated that external finance would be necessary to modernise telecommunications in Britain, and various financial schemes were considered until outright divesture was finally chosen (Parker, 2009). Due to the Conservative Party's desire to reduce the monopoly power of British Telecom (Parker, 2009, p. 244), initial efforts were made to split it up prior to the sale, citing the break-up of AT&T in the United States as a model. The unions, some managers, domestic equipment suppliers, the Labour Party and significant public opinion were against privatisation *per se*, whilst unions and managers alike strongly opposed splitting up the enterprise (Parker, 2009, p. 295). Privatisation would be more palatable all round if British Telecom was sold as a single entity. For political reasons, privatisation was delayed until after the 1983 election. Meanwhile, the government settled with the second best option: forging limited competition (Florio, 2003). With the British Telecommunications Act of 1981, the liberalisation of apparatus and maintenance was introduced. Simultaneously, the government helped engineer the establishment of a rival to British Telecom, Mercury Communications, owned initially by a consortium of Cable & Wireless, BT and Barclays Bank. Soon afterwards, both BP and Barclays withdrew leaving Cable & Wireless the sole owner. To protect and encourage Mercury, it was licensed as British Telecom's only competitor as part of the 'duopoly' policy. Members of the government actively encouraged Mercury to engage in the broadest competition possible, but the firm refused to build a national network, preferring to use existing lines to cream-skim profitable business (Parker, 2009, p. 246).

In 1984, the initial public offering of 51% of British Telecom shares was sold. This was the first time Britain had sold a national monopoly, and the sale was one of the largest public offerings to date, requiring the establishment of a new regulatory approach (known as the RPI-X formula) and an independent regulatory agency, the Office for Telecommunications. The way in which British Telecom was privatised came to represent a blueprint for the sale of other utilities, in Britain, Europe, Latin America and beyond (Clifton, 2000). The reasons behind the Conservative Party's decision to privatise British Telecom in 1984 can be summarised into three main points: firstly, from the practical side, it was recognised that new technology would require levels of investment that the government, constrained by its public debt, would not provide. As a private firm with growth potential, attracting this finance would not be an issue. Secondly, the government's ideological stance was in favour of reducing the monopoly power of British Telecom and its unions, especially the Post Office Engineering Union, via the introduction of competition (Parker, 2009, pp. 241–244). Thirdly, from a political economy perspective, the government was lobbied by the City and the Bank of England who claimed London's hub as an international financial centre would be at risk without modern and efficiently working communications services.

The sale was subject to much publicity including television, press, poster, radio, leaflet, and travelling road show campaigns. The offer was 3.2 times over-subscribed and concluded with 2.15 million investors (Parker, 2009, p. 312). Nigel Lawson, then Chancellor of the Exchequer, lauded the sale as 'the birth of people's capitalism' (Lawson, 1992, cited in Parker, 2009, p. 320). British Telecom was finally working under private company law, freed from former borrowing restrictions. The 1991 White Paper ended the duopoly, opening the door to increased competition from domestic and foreign providers. Renamed BT, the old red telephone boxes were replaced with glass ones painted with the Pied Piper, whilst the firm was reorganised into profit centres, and staff retrained along commercial lines. Corporate governance was reformed, rendering its directors much more accountable towards shareholders than was the case in its Italian, French and German peers (Goldstein, 2000, p. 203), including through generous bonuses. BT helped facilitate the decline of the power of its unions (Hulsink, 1999, p. 155), which in turn facilitated the introduction of more flexible labour contracts, merit-based promotion and less generous pension schemes. Total privatisation was completed in stages: in 1991, the government sold off half of its 47.6% of shares, and two years later, virtually all of its remaining shares. Finally, the incoming Labour government sold off its 'golden share' in 1997 which had been retained initially to avoid take-overs (BT, undated).

BT abroad

BT's new business environment, particularly technological and regulatory change, plus its new financial freedom, opened the door to its internationalisation. Two main routes to internationalisation were through FDI and global alliances. BT opted largely for global alliances, particularly with large players based in the United States. Though its internationalisation project started ambitiously, with hindsight it has been widely regarded as a failure (Turner & Gardiner, 2007). In 1994, BT (71.5%) and the United States' second largest long-distance telecommunications firm, MCI (24.9%), entered a joint venture to provide global communication services to multinational corporations by establishing Concert. By 1996, after a number of other operators, including Telefónica, joined, Concert was positioned to emerge as the leading provider of its kind, however, just before finalising the merger between BT and MCI, MCI was bought by WorldCom. Telefónica abandoned the alliance to join WorldCom-MCI. Again, in 1998, BT and AT&T agreed to form a joint venture to serve multinational corporations, but the new Concert project was abandoned in 2001, and the companies were forced to embark on a costly de-merger. BT was also forced to de-merge from its mobile activities in the form of MMO2, to reduce its debt caused by acquisitions and 3G licenses (van Kranenburg & Hagedoorn, 2008, p. 122).

Conclusions

This article sought to explain why, after regulatory change across the telecommunications sector from the 1970s, some of Europe's former national telecommunications monopolies embarked on vigorous and successful international ventures, joining the ranks of the world's leading multinational corporations, whilst others were much less successful, internationalising either hesitantly, or else going abroad, failing and later de-internationalising. For some scholars of international business, this uneven road to internationalisation could be explained by short-term regulatory asymmetries between countries (Sarkar et al., 1999; Bonardi, 2004). That is, former monopolies based in countries where liberalisation and privatisation were delayed or restricted, relative to elsewhere, would internationalise more strongly, enjoying, as they did, greater access to monopoly rents, political diplomacy and inside information to facilitate their adventures abroad. Elsewhere, the authors showed this explanation to be unsatisfactory, since no significant correlation was found between firm internationalisation, and speed and depth of liberalisation and privatisation (Clifton et al., 2010). Data on firm internationalisation, liberalisation and privatisation presented in this article confirmed that firms internationalised strongly both from country bases where policy had reached different stages. So, explanations for different internationalisation must be found at the firm level. Taking as the point of departure the notion that firms do not operate in a vacuum, but are embedded in national contexts, including different regulatory frameworks, the article continued by exploring how regulation might have influenced firms' later position and approach to the opportunity to internationalise. Following 'maximum variation' case study methodology (Flyvbjerg, 2006), two large telecommunications firms were selected for analysis: Telefónica and BT. Both were based in large countries where privatisation and liberalisation had either been implemented earlier or on time, with respect to the European average but, these firms embodied two highly differing – near opposite – internationalisation strategies. Telefónica ventured abroad very early on, in a vigorous and unrelenting fashion, mainly through FDI, emerging as one of the worlds' most international of telecommunications former monopolies (UNCTAD, 2010). BT started to internationalise somewhat later, though it was more active compared with many of its European peers. However, global alliances were prioritised over FDI and, each of these alliances failed successively. Later on, BT managers decided to partly de-internationalise and refocus on its home market. BT became the least international of the large European players.

Explanations for radically different paths to internationalisation can be found in an analysis of the historical evolution of the different regulatory frameworks in which each incumbent developed, particularly, through ownership, management style, access to capital and exposure to liberalisation. As regards ownership, Telefónica was exceptional in the European context in that it was born not only as a private company, but as one controlled by an emerging US multinational corporation, ITT. Though private, foreign influence was greatly reduced when the firm was nationalised in 1944 under the Franco regime, nationalisation was never total since the firm always remained partly controlled by private national banks and multiple shareholders. As regards management, Telefónica was never run by a PTT; management style was closer to that of a private firm than public administration. Moreover, even from its origins in 1924, Telefónica accumulated experience in generating finance externally, issuing multiple public offerings, even though these were subject to limits due to partial state ownership. From the 1980s, much of this finance would be obtained in the international market and channelled towards its corporate internationalisation, so these operations did not strain government budgets. The challenge of technological change in the telecommunications sector was met in Spain by a Socialist government in an optimistic period of democratic transition and membership of the European Community. As part of a general project of industrial restructuring, Telefónica was forced to change its management style. It broke its technological dependence on Standard Eléctrica (ITT) while forging technological and business alliances with suppliers world-wide, and improving domestic R&D. It was partially privatised, embarking on an internationalisation strategy using finance from public offerings. Despite the fact that Spain was given a formal extension by the European Commission to introduce telecommunications liberalisation (as Portugal, Ireland and Greece) it chose to assume the earlier deadline of 1998, in line with most other European countries. Nevertheless, Telefónica, unlike British Telecom, did enjoy some pampering as a 'national champion'. Importantly, Telefónica's foreign acquisitions in Latin America in the aftermath of the debt crisis were conducted from the cosy base of monopoly status. Despite formal liberalisation, Telefónica continued to enjoy a high market share vis-à-vis many of its European peers.

The ownership and management style of telecommunications in Britain contrasted sharply with the experience of Telefónica. In Britain, telecommunications were consolidated as a national network under state ownership and public administrative control from the end of the nineteenth century. Despite regular criticism that its managerial style was overly bureaucratic from the 1930 to the early 1960s, no major changes were introduced. As an arm of government, investment in British telecommunications was strictly limited, whilst its managers lacked commercial vision (Lipartito, 2004). Technological changes in the communications sector was a common watershed for governments around the world. In Britain, the government and the City concluded that the required investment could be generated were British Telecom to be privatised. This was used to justify its privatisation, from 1984, though the government also had ideological reasons to promote its sale, particularly, to weaken the trade unions and the monopoly power of the firm (Parker, 2009), a policy later emulated by Latin American governments (Clifton, 2000). Though some in the government wished to split up British Telecom prior to the sale, when this failed, an alternative route was taken: the government would support the emergence of a rival, Mercury. The British government started to liberalise telecommunications markets well ahead of the European directive. Despite the fact that competition pressures on BT were limited (Florio, 2003), they were still greater than those faced by most of its European peers. In so doing, the government revealed they were not prepared to promote British Telecom as a 'national champion'. When mobile phone licences were auctioned in the 1980s, Racal Electronics and Cable & Wireless, renamed Vodafone, emerged. Its international operations would soon overshadow the incumbent. With an increasingly contested home market and, without 'national champion' backing, BT opted to internationalise through global alliances, particularly with US firms such as MCI. This strategy was generally perceived as a quicker and less risky path towards internationalisation than that of FDI (Ulset, 2008). When its alliances failed successively, it is hardly surprising that management decided to refocus efforts on protecting its market share at home, effectively, partially de-nationalising.

Notes

1. Telefónica was originally called the Compañía Telefónica Nacional de España until the 1980s, when, in a re-branding exercise, this was shortened to Telefónica. For convenience, Telefónica will be used here.
2. Detailed results of the correlation analysis can be found in more detail in Clifton et al. (2010).
3. Little (1979) describes the 'wheeling and dealing' between the Behn brothers and highly-placed Spanish officials and US Ambassadors to favour ITT's successful control of the telecommunications market.
4. ITT was considering selling off Standard Eléctrica (*El País*, 8 January 1981), and did so in 1988 to Alcatel NV, a subsidiary of the French publicly owned multinational Compagnie Générale d'Électricité.
5. The incumbent supplier of telecommunications in Britain was called British Telecom from 1980 and BT from 1991.

References

Álvaro, A. (2007). Redes empresariales, inversión directa extranjera y monopolio: el caso de Telefónica, 1924–1965. *Revista de Historia Industrial, 34*, 65–96.

Álvaro, A. (2008). Changing patterns of ownership and labour management under changing institutional regimes: The case of Telefónica in Spain, 1924–1970. In S. Fellman, A. Kuustera, & E. Vaara (Eds.), *Historical perspectives on corporate governance: Reflections on ownership, participation and different modes of organising* (pp. 243–259). Helsinki: Finnish Society of Science and Letters.

Antonelli, C. (1995). Localized technological change in the network of networks: The interaction between regulation and the evolution of technology in telecommunications. *Industrial and Corporate Change, 4*, 737–754.

Bauer, J. (2010). The changing role of the state in telecommunications. *International Telecommunications Policy Review, 17*, 127–145.

Bel, G., & Trillas, F. (2005). Privatization, corporate control and regulatory reform: The case of Telefónica. *Telecommunications Policy, 29*, 25–51. doi: 10.1016/j.tel.pol.2004.09.003

Bonardi, J.P. (2004). Global and political strategies in deregulated industries: The asymmetric behaviors of former monopolies. *Strategic Management Journal, 25*, 101–120. doi: 10.1002/SMJ.367

BT. (undated). Events in telecommunications history, 1984. Retrieved from http://www.btplc.com/Thegroup/BTsHistory/1984onwards/1984.htm

Bureau Van Dijk. (undated). *Amadeus database.* Retrieved from https://amadeus.bvdinfo.com/version-20101125/home.serv?product=amadeusneo&loginfromcontext=ipaddress

Calvo, Á. (2007). Redes empresariales, inversión directa extranjera y monopolio: El caso de Telefónica 1924–62. *Revista de Historia Industrial, 34*, 59–81.

Calvo, Á. (2008). State, firms and technology. The rise of multinational telecommunications companies: ITT and the Compañía Telefónica Nacional de España, 1924–1945. *Business History, 50*, 455–473.

Calvo, Á. (2010). *Historia de Telefónica: 1924–1975. Primeras décadas:tecnología, economía y política*. Barcelona: Fundación Telefónica-Ariel.

Chari, A., & Gupta, N. (2008). Incumbents and protectionism: The political economy of foreign entry liberalization. *Journal of Financial Economics*, *88*, 633–656.

Clifton, J. (2000). One the political consequences of privatisation: The case of Teléfonos de México. *Bulletin of Latin American Studies*, *19*, 63–79. doi: 10.1111/j.1470-9856.2000.tb00092.x

Clifton, J., Comín, F., & Díaz-Fuentes, D. (2003). *Privatization in the European Union: Public enterprises and integration*. Dordrecht: Kluwer Academic.

Clifton, J., Comín, F., & Díaz-Fuentes, D. (2007). *Transforming public enterprise in Europe and North America. Networks, integration and transnationalization*. Basingstoke: Palgrave Macmillan.

Clifton, J., Comín, F., & Díaz-Fuentes, D. (2008). The rise of the new public service transnationals. In H. Schroeter (Ed.), *The European enterprise* (pp. 209–222). Dordrecht: Springer.

Clifton, J., & Díaz-Fuentes, D. (2010a). Evaluating EU policies on public services: A citizens' approach. *Annals of Public and Cooperative Economics*, *81*, 281–311. doi: 10.1111/j.1467-8292.2010.00412.x

Clifton, J., & Díaz-Fuentes, D. (2010b). Is the European Union ready for FDI from emerging markets? In K. Sauvant, W.A. Maschek, & G. McAllister, *Foreign direct investment in emerging markets: The challenges ahead* (pp. 335–358). New York: Palgrave.

Clifton, J., Díaz-Fuentes, D., & Revuelta, J. (2010). The political economy of telecoms and electricity internationalization in the single market. *Journal of European Public Policy*, *17*, 988–1006. doi: 10.1080/13501763.2010.499229

Comín, F. (2008). Public enterprises in Spain: Historical cycles and privatizations. *Análise Social*, *43*, 693–720.

Comín, F., & Díaz-Fuentes, D. (2004). *La empresa pública en Europa*. Madrid: Síntesis.

Comín, F., & Gálvez, L. (2006). Reglas del juego, redes de oportunidad y nacionalizaciones de empresas extranjeras durante el franquismo. In J. Tascón (Ed.), *Redes de empresas en España. Una perspectiva teórica, histórica y global* (pp. 115–133). Madrid: Lid.

Committee of Enquiry on the Post Office. (1932). *Report of the Committee of Enquiry on the Post Office, Cmd 4149*. London: HMSO. Retrieved from http://www.btplc.com/Thegroup/BTsHistory/Eventsintelecommunicationshistory/Eventsintelecommunicationshistory.htm

Dunning, J. (2009). Preface. In K. Sauvant & L. Sachs (Eds.), *The effect of treaties on foreign direct investment* (pp. xxv–xxvi). Oxford: Oxford University Press.

El negocio de ITT en España. (1983, January 1). *El País*. Retrieved from http://www.elpais.com/articulo/economia/TELEFONICA/CITESA/ITT/negocio/ITT/Espana/elpepieco/19831101elpepieco_9/Tes

Florio, M. (2003). Does privatisation matter? The long-term performance of British Telecom over 40 years. *Fiscal Studies*, *24*, 197–234.

Flyvbjerg, B. (2006). Five misunderstandings about case-study research. *Qualitative Inquiry*, *12*, 219–245.

Gálvez, L., & Comín, F. (2003). Multinacionales, atraso tecnológico y marco institucional. Las nacionalizaciones de empresas durante la autarquía franquista. *Cuadernos de Economía y Dirección de la Empresa*, *17*, 139–179.

Goldstein, A. (2000). Corporate governance and regulation in privatized utilities: Telecommunications in four European countries. *Business and Politics*, *2*, 189–223. doi: 10.1080/1369525002000646 1

Guillén, M. (2005). *The rise of Spanish multinationals. European business in the global economy*. Cambridge: Cambridge University Press.

Haar, L., & Jones, T. (2008). Misreading privatization and liberalization: The case of US energy utilities in Europe. *Energy Policy*, *36*, 2610–2619.

Hertner, P., Hausman, W., & Wilkins, M. (2009). Global electrification. *Multinational enterprise and international finance in the history of light and power*, 1878–2007. Cambridge: Cambridge University Press.

Hulsink, W. (1999). *Privatisation and liberalisation in European telecommunications*. London: Routledge.

Ledbetter, R. (1985). ITT: A multinational corporation in Latin America during World War II. *The Historian*, *47*, 7.

Lipartito, K. (2004). Failure to communicate: British telecommunications and the American model. In J. Zeitlin & G. Herrigel (Eds.), *Americanization and its limits* (pp. 153–179). Oxford: Oxford University Press.

Little, D. (1979). Twenty years of turmoil: ITT, the State Department, and Spain, 1924–1944. *Business History Review*, *53*, 449–472.

López, S. (2003). The role of Telefónica: The internationalisation of telecommunications in Spain, 1970–2000. *Business and Economic History On-Line*, *1*. Retrieved from http://www.h-net.org/~business/bhcweb/publications/BEHonline/2003/Lopez.pdf

Millward, R. (2005). *Public and private enterprise in Europe: Energy, telecommunications and transport*. Cambridge: Cambridge University Press.

Millward, R. (2007). Cross-border investment and service flows in networks within Western Europe. In J. Clifton, F. Comín, & D. Díaz-Fuentes (Eds.), *Transforming public enterprise in Europe and North America* (pp. 16–29). London: Palgrave.

Millward, R. (2010, June). *Public enterprise in the modern western world*. Paper delivered at the Milan European Economy Workshop, Milan Italy.

Noam, E. (1992). *Telecommunications in Europe*. Oxford: Oxford University Press.

OECD. (undated). *Product Market Database, 1970–2008*. Retrieved from http://stats.oecd.org/Index.aspx?DataSetCode = PMR

OECD. (1999). *Communications outlook*. Paris: OECD.

OECD. (2001). *Communications outlook*. Paris: OECD.

OECD. (2003). *Communications outlook*. Paris: OECD.

OECD. (2005). *Communications outlook*. Paris: OECD.

OECD. (2008). *Communications outlook*. Paris: OECD.

OECD. (2009). *Communications outlook*. Paris: OECD.

Parker, D. (2009). *The official history of privatisation in Britain. Volume 1*. London: Routledge.

Pelkmans, J. (2001). Making EU network markets more competitive. *Oxford Review of Economic Policy*, *17*, 432–456. doi: 10.1093/oxrep/17.3.432

Puig, N., & Álvaro, A. (2005). La guerra fría y los empresarios españoles: La articulación de los intereses económicos de Estados Unidos en España, 1950–1975. *Revista de Historia Económica*, *22*, 385–424.

Ramamurti, R. (Ed.). (1996). *Privatizing monopolies: Lessons from the telecommunications and transport sectors in Latin America*. Baltimore, MD: Johns Hopkins University Press.

Rodriguez, M. (2005, April 28). Las 'Matildes' celebran ocho décadas de capitalismo popular. *Cinco Días*. Retrieved from http://www.cincodias.com/articulo/mercados/matildes-celebran-decadas-ncapitalismo-popular/20050428cdscdimer_4/cdsmer/

Sarkar, M.B., Cavusgil, S., & Aluakh, P. (1999). International expansion of telecommunications carriers: The influence of market structure, network characteristics and entry imperfections. *Journal of International Business Studies*, *30*, 361–382. doi: 10.1057/palgrave.jibs.8490074

Schröter, H. (2005). *Americanization of the european economy*. Dordrecht: Springer.

Sobel, R. (2000). *ITT: The management of opportunity*. Washington, DC: Beard Books.

Solana, L. (1979, October 17). El grupo ITT en España, hoy. *El País*. Retrieved from http://www.elpais.com/articulo/economia/ESPANA/STANDARD_ELECTRICA/TELEFONICA/MARCONI/CITESA/ITT/grupo/ITT/Espana/hoy/elpepieco/1979107elpepieco_7/Tes

Tascón Fernández, J. (2005). La red yanqui desde los años treinta hasta los años del milagro. In J. Tascón Fernández (Ed.), *Redes de empresas en España. Una perspectiva teórica, histórica y global*, pp. 137–154. Madrid: Lid.

Turner, C., & Gardiner, P. (2007). De-internationalisation and global strategy: The case of British Telecommunications (BT). *Journal of Business and Industrial Marketing*, *22*, 489 – 497. doi: 10.1108/08858620710828863

Ulset, S. (2008). The rise and fall of global network alliances. *Industrial and Corporate Change*, *17*, 267–300. doi: 10.1093/icc/dtn003

UNCTAD. (2004). *World investment report. The shift towards services*. Geneva: UNCTAD.

UNCTAD. (2008). *World investment report. Transnational corporations and the infrastructure challenge*. Geneva: UNCTAD.

UNCTAD. (2010). *World investment report*. Geneva: UNCTAD.

van Kranenburg, H.L., & Hagedoorn, J. (2008). Strategic focus of incumbents in the European telecommunications industry: The cases of BT, Deutsche Telekom and KPN. *Telecommunications Policy*, *32*(2), 116–130.

Wilkins, M. (1974). *The maturing of multinational enterprise: American business abroad from 1914 to 1970*. Cambridge, MA: Harvard University Press.

Wilkins, M. (2004). *The history of foreign investment in the United States 1914–1945*. Cambridge, MA: Harvard University Press.

Zeitlin, J., & Herrigel, G. (Eds.). (2004). *Americanization and its limits*. Oxford: Oxford University Press.

Public utilities and private initiative: The French concession model in historical perspective

Dominique Barjot

Department of History, Paris-Sorbonne (Paris IV) University, Paris, France

In France, the Public-Private Partnership allowed the conciliation of the social function of public services and public works and the limitation of their cost for each citizen. They were in three sectors: transport, energy and water, during the nineteenth century and the first part of the twentieth century.

After the Second World War, in spite of nationalisations, the concession system continued to operate in water treatment and distribution, but also in transport and energy. Moreover, the necessity to modernise French infrastructures favoured a revival of the concession system combined with Anglo-Saxon practices.

The French model of the concession constituted a major asset for French capitalism, with big multinational firms such as GDF-Suez, Veolia Environnement, Vinci and Bouygues.

1. Introduction

In France, the economic influence of the state was and remains strong (Barjot, 2009a; Toninelli, 2000). Over the long term, Public-Private Partnerships constituted a means of conciliating the social function of public services and public works, whilst limiting their costs for citizens. The nineteenth and twentieth centuries offer a number of examples of the use of the concession system in the utilities sector, including the transportation, energy and water sectors. During the nineteenth century, railways, firstly, the main network, supported by national interests, then the other lines, advocated by local interests, were built with the support of important financial groups. Later on, coal mines, gas and electricity were developed with contributions from large financial groups according to the same model. Despite the nationalisation of railways (1937), coal mines (1944 and 1946) and electricity and gas (1946), the concession system remained in operation (Barjot, 2006b). It was always used in water treatment and distribution, even when this sector was fully nationalised. The need to develop, then modernise, the French highway networks (from 1969) and build expensive public works (such as the Eurotunnel or the Stade de France), numerous urban services, such as underground car parks, or administrative buildings, such as prisons and universities, saw a return to the concession system, which was combined with Anglo-Saxon practices, including Build Operate Transfers (BOT) and Public-Private Partnerships, or, PPP (Perrot & Chatelus, 2000). Moreover, the French model of concession contributed significantly towards the international development of French capitalism across the nineteenth century, such as the building of the Compagnie du Canal de Suez (Piquet, 2008). Other large concessions were crucial for the development of the French Empire and its colonies. After the Second World War, the concession remained a competitive system, providing the framework for the subsequent rise of French multinational groups, particularly from the 1990s, such as EDF, GDF-Suez, Veolia Environnement and Vinci and Bouygues. The main argument of this article is that the French legal system stimulated 'network capitalism' by establishing a financial and managerial system which allowed for the participation of private capital in the development of major infrastructure projects. This, in turn, gave French companies a solid advantage when they internationalised, most particularly, from the 1990s. The article first explains the origins and significance of this legal system before exploring how this facilitated the rise of French multinationals in different branches of infrastructure.

2. The French concession system

The concession system in France was historically characterised by a number of specificities, particularly from the legal point of view (Martinand, 1993). In the recent period, the concession system is undergoing a revival (Petitet & Varaschin, 1999). This new form of concession is referred to as the delegation of management (Bettinger, 1997).

2.1. Definitions: concession of public utilities and public works

There are two groups of concession in the French model: the concession of public utilities – '*concession de service public*' – and the concession of public works – '*concession de travaux publics*' – as stated by Bezançon (1999).

A. Public utilities

In the French case, the most significant difficulty comes from the existence of a specific branch of law, administrative law (Chevallier, 1987; Mescheriakoff, 1991). According to this branch of law, there are two definitions of public utility. From a material point of view, public utilities correspond to a set of human and material means established to realise an administrative task. From the jurisdictional point of view, however, public utilities refer to an activity of the general interest managed or controlled by the public administration but associated with a regime that falls totally or partially under private law. As a consequence, a private entity can deliver public services. From the beginning of the twentieth century, French administrative jurisprudence has allowed for public utilities to be managed by a private entity.

There are four major ways to manage public utilities: to lease them to an entity which does not invest in them but simply manages them; to establish a contract with an entity which receives payment for services; direct management by the administration itself or by an agent; or, via the concession. Of these four options, the last two are the most common. In the case of direct administration, public utilities are not financed by a specific budget and are not managed to make profit. In both cases, the administration is responsible for covering any deficits.

Under the concession system, a public entity entrusts a private entity, the concessionaire, to manage public utilities with remuneration, according to a public contract. This contract defines statutory and contractual specifications. From a material point of view, the concession contract includes two documents: a convention, which explains the general aspects of the service; and a specifications book, which gives details on the organisation and functioning of the service.

Public utilities under the concession regime remain under the public domain. In the contemporary period, a concession is generally granted by open or limited tenders but, during the mid-1950s, the most common system was to allocate the concession to the lowest tender. Nevertheless, in the case of urgent needs or due to technical difficulties, the administration could proceed over-the-counter, by means of competitive bidding. Consequently, the concessionaire can obtain an indemnification if the conceding administration decides unilaterally on a change in the carrying out of the contract, or a compensation for a change in the economic agreement and, if this is impossible, a new contract can be negotiated.

The concession ends with the termination of the contract. French administrative justice does not accept renewal by tacit agreement. But the concession can finish before the term comes to an end. The conceding administration can repurchase it by forfeiture if the concessionaire does not respect the statutory or contractual rules, if the general interest disappears, or if that decision is taken by the administrative justice system.

B. Public works

The French administration has different methods by which it executes public works. One means is direct management: in this case, the administration realises the work using its own workforce and materials. A second method consists of using a contractor, via a public works contract. Here, the contractor executes the works on the payment of a fixed fee. A third way is the concession: a concession contract implies that a private entity will execute the public work, such as construction, or the supply of equipment.

2.2. An old tradition which leads to complex practices

The concession tradition goes back approximately two millennia (Bezançon, 2004a). It was further developed during the Ancient Régime period, and reached its peak during the nineteenth century. According to Xavier Bezançon (2005), the importance of the concession in French history grew, declined, to be later revived once again. During the Roman period, the law of 'public orders' was in force: this became the basis for concession contracts. After a period of decline, the rediscovery of Roman law allowed for a revival of concessionary practices. During the seventeenth and eighteenth centuries, concessions were used to facilitate the building of canals and, later, bridges (Conchon, 2005). In the aftermath of the French Revolution, the State refused to define concession by law. Consequently, during the French First Empire, the Restoration, the Monarchy of July and the Second Empire, the concession regime was only defined by the jurisprudence of the State Council (*Conseil d'Etat*). Facing the dominant rule of free enterprise, the State Council established the concept of 'public utilities' ('public service'), which became increasingly influential in the development of canals, railways, maritime transportation (Barjot & Berneron-Couvenhes, 2005; Berneron-Couvenhes, 2007), the telegraph and, later, electricity, petrol and aeronautics (Marnot, 2005).

The high degree of complexity of problems led to the emergence of equally complex solutions (Barjot, Petitet, & Varaschin, 2002). One reason for this was the development of new networks and services at the national and local levels and, at the same time, the acquisition of new competencies by departments and communes, according to the laws of 1871 and 1884. During the 1880s, across Western Europe in general and, particularly in France, the municipal socialism movement emerged. With the First World War and the onset of strong and chronic inflation, the attitude of the state (and, to a certain degree, departments and communes), changed in favour of direct administration and the constitution of public enterprises, such as SNCF (1937), then EDF and GDF, or, sometimes, of a mixed model, CNR, 1921 (see Barjot, Morsel, & Badel, 1996). By the end of the 1930s, the concession presented interesting solutions to public budget deficits. Consequently, the concession system developed in two dominant sectors: public works on the one hand (Barjot, 2006a), and public utilities, on the other (Bettinger & Le Chatelier, 1995).

3. The first privileged sector: public works

During the nineteenth century, the concession of public works was very common (Barjot, 1989; Reverdy, 2004). This was largely because the administration preferred not to execute the largest basic investments directly. This became particularly evident with the failure of the Freycinet Plan (Barjot, 1993), after which the concession was broadly adopted. Nevertheless, as the concessionaire had to build, and then exploit, the public works, concessions were increasingly connected to the concession of public utilities, though a distinction remained between the two concepts.

3.1. The concession of public works: from decline to renewal

During the first part of the twentieth century, this situation changed, because the notion of the public utility concession came to progressively absorb that of public works concession.

A. The concession: an efficient means to optimise public expenditure

The concession of public works was revived after the Second World War (Barjot & Berneron-Couvenhes, 2005) and the state used this method to build and exploit multiple public works. With the Act of 18 April 1955, the state conceded the building and the exploitation of highways to public organisms and companies which were financed by a mix of both public and private capital, which would be authorised to collect tolls. Following the Act of 12 May 1970, the state permitted private firms to be recipients of concessions for highways. The same solutions (Barjot, 1990) were adopted on building the Tancarville Bridge over the Seine River (Act of 17 May 1951), and on excavating the tunnel under Mont Blanc between 1965 and 1971 (Berthereau, 2005b). The administration chose a concessionaire based either in France or in another European Community member state, though the concessionaire had to apply the rules defined by the French administration (such as the tariffs to be paid by users). In return, the concessionaire had the right to compensation where these rules did not allow for a healthy financial outcome of the concession. The concessionaire had other rights, including the right to expropriate, and to benefit from the privilege of exploitation. In contrast, the conceding administration can exert a right of control, though not to directly manage the service (Berthereau, 2005a). The conceding administration has the power to impose a one-sided change and disciplinary action (foreclosure, or even forfeiture) from the concessionary.

In the more recent period, the concession of public works has been adopted to realise many major public works, both in France and abroad. Overseas projects included the construction of two bridges across the Humber River in the UK, the confederation bridge in Canada, and the Vasco de Gama on the Tage River (Barjot, 2003a). In France, this method was used to construct the Stadium of France (Stade de France) at Saint-Denis, and the viaduct of Millau (Barjot, 2008; Barjot & Dureuil, 2008). So, the concession constitutes an efficient means to optimise public expenditures, particularly for urban public utilities, whilst it also constitutes a useful instrument for development in emerging countries.

B. The concession as an instrument of growth in developed and developing countries

The concession was the result of an old tradition of French public works abroad particularly between the middle of the nineteenth century and the beginning of the twentieth century (Barjot, Petitet, & Varaschin, 2002). From the Second Empire, the Rothschild family and the Pereire brothers competed to obtain the biggest concessions of railway networks. At the same time, Ferdinand de Lesseps built, again, under the regime of the concession, the Suez canal, between 1854 and 1869 (Montel, 1998; Piquet, 2009). The concession system was also used to provide equipment to the French colonial Empire. French contractors, such as the *Société de Construction des Batignolles* (SCB) at Tunis (Park, 2000), or Hildevert Hersent at Bizerte or Dakar, built harbors and railways (Barjot, 1985). Later, at the beginning of twentieth century (Park, 2005), many French contractors internationalised strongly thanks to the concession system (Barjot, 1995).

An important turning point was clinching, in 1902, the concession of the Rosario harbour by Hildevert Hersent and Henri Schneider, which was the main center of wheat exportation from the Argentine Pampa to Europe (Barjot, 2003c). From this date onwards in South America, most port infrastructure would be built under the concession system by French contractors. This was also the case in China with regard to railways (though generally in cooperation with Belgian capitalists and entrepreneurs), as well as in Russia, in the cases of ports, railways and, also electricity.

This also happened in Western Europe. One good example was the concession of the Port of Lisbon to the Hersent brothers, Jean and Georges, from 1899 to 1921 (Barjot, 1994) French contractors undertook important projects abroad. For instance, the Société des Grands Travaux de Marseille, Société Générale d'Entreprises and the Fougerolle brothers undertook work across the Russian and Ottoman Empire. Indeed, the Mediterranean basin became an area of privileged development for French concessionaires, including Egypt (Saul, 1997), Syria, and Greece, where the SCB built the Piraeus-Salonika railway line under the concession regime. Supporting the connection between the Greek and Ottoman networks, the Société des Chemins de Fer helléniques, concessionaire of the line, played a decisive role during the First World War to the benefit of Allied countries. As was often the case, this concession benefited developing countries.

3.2. Two contemporary French world leaders: Vinci and Bouygues

From the middle of the 1980s, French firms emerged as world leaders in the civil engineering sector as well as the construction industry. This remained so until 2009, when two Chinese firms, China Railways Construction and China Railways Group, respectively, emerged as the first and the second most important players in the world market (New Engineering Records, 2009). Bouygues had became world leader in 1987, after its merger with SCREG (Barjot, 2010); in 2004, Vinci succeeded Bouygues as number one. In 2009, Vinci ranked as the third most important player, followed by Bouygues in fourth position.

A. Vinci

From 2000, the Vinci Group rose to be the world leader in the construction and public works sectors (Barjot, 2003a). The Vinci Group was the outcome of the merger between the two most prestigious French civil engineering firms: Société Générale d'Entreprises (SGE), set up in 1899, and Société des Grands Travaux de Marseille (GTM), founded in 1891. After the First World War, the two firms had often been rivals, but also, frequently, partners. Even though they were clearly subject to individual path-dependency processes, they converged towards common models of corporate governance, ownership structure and industrial relations. This in fact facilitated their subsequent merger – or, more accurately, takeover – of SGE of GTM, creating Vinci, the world leader.

In 1945, SGE was one of the leading civil engineering firms in France and the mother company of the fourth most important French electricity company after the Mercier, Durand and Empain groups. The nationalisation of electrical power industry deprived the firm, in 1946, of its more prosperous subsidiaries. SGE accepted the challenge, becoming, between 1946 and 1966, the French construction leader and the third most important player within the European Community, thanks a strategy of internal and external growth. It became a subsidiary of the Compagnie Générale d'Electricité from 1967 to 1983, then Saint-Gobain, between 1983 and 1987. SGE was reinforced by a succession of mergers, with Sainrapt et Brice (1979), SOBEA, the most important subsidiary of Saint-Gobain in the construction sector (1986), Campenon Bernard, the world leader of the pre-stressed concrete (1988), and a number of take-over operations. By 1999, the new SGE Group stood as a European leader in the construction and civil engineering field, with strong positions in Belgium, Germany, the United Kingdom and Eastern Europe.

GTM was the French leader in the construction sector between 1907 and 1939. In 1945, GTM could not bank on the same goodwill as SGE had enjoyed, having carried out some activities in collaboration with German firms during the First World War. Furthermore, despite the firm's links with the Energie Electrique du Littoral Méditerranéen, subsidiary of the Compagnie Française Thomson-Houston, GTM did not enjoy the same advantages that SGE had in the electrical power sector. At the same time, GTM contributed substantially to the reconstruction of post-war France. More export-oriented than SGE, GTM set up a number of lasting overseas operations in countries such as Argentina, Brazil and Canada. At the same time, the firm developed successfully its pre-stressed concrete activity, and made a breakthrough in the nuclear power sector and in off-shore facilities, with the setting up by GTM and Entrepose of a major joint venture subsidiary,

Entrepose pour les Travaux Publics et Maritimes (ETPM). Transformed in 1968 into a holding company, GTM played a pioneering role in the area of operating concessions for public works such as underground car parks and motorway concessions, founding Cofiroute, through an alliance with the Entreprise Jean Lefebvre (EJL), number three in the French road works industry. In 1981, facing the second oil crisis, GTM and Entrepose merged into a new corporation: GTM Entrepose.

In 1986, because of unsatisfactory technological performance, GTM Entrepose came under the control of Dumez, one of its most important competitors. French exporter in the civil engineering sector, Dumez merged, in 1990, with Lyonnaise des Eaux, to form Lyonnaise des Eaux-Dumez. Subsequently, GTM-Entrepose became a subsidiary of the new group. Then, because of the Dumez Group's chaotic diversification and its consecutive losses, GTM soon regained its independence. In 1993, Dumez itself became a subsidiary of GTM Entrepose, then Group GTM (1998). GTM now divested itself of its real estate and oil industry activities and asserted its leadership in the operation of concessions: underground parks, roads and motorways, large public-works structures (the Rion Antirion bridge for instance) and airports (Beijing).

One of the major stakes of the merger was the separation from the two parent companies: Compagnie Générale des Eaux (then Vivendi), from SGE, the firm that had begun the takeover, and Suez Lyonnaise des Eaux, from GTM, as the firm had been taken over. This was a success: from 2001, the Vinci Group was controlled through a broad international ownership. Whilst not forsaking the construction sector, the group largely switched to public works, and started activity as a service provider, building underground car parks, motorways, harbours and airports (especially in China and Mexico), but also public works (Stade de France in Saint-Denis, the Humber Bridge, Confederation bridge and so on). Reinforced by successful operations of external growth (Solétanche, Taylor Woodrow Construction, Entrepose Contracting, Cégelec and, in 2006, Autoroutes du Sud de la France), by 2009, the group employed some 162,000 staff across about 100 countries (while Bouygues employed 134,000 across around 90 countries). Today, Vinci finances, designs, build and manages facilities. In 2009, concessions constituted the most profitable sector of activities of Vinci.

In 2009 (see Table 1), concessions were mainly realised in France (94% of revenue), with some 2% generated elsewhere in Europe and 4% in the rest of the world. On this basis, Vinci made an operating profit rate of 9.8% (10.1% in 2007), which was higher than that of Bouygues, its most important domestic (and European) competitor (which made 5.9%).

Table 1. Financial performance of Vinci in 2009 (in %).

	Concessions	Contracting	Holding companies
Revenues	15	85	
Operating profits from ordinary activities	60	38	3
Cash flow from operations	62	35	3
Net profit attributable to equity holders	47	50	3

Source: *Vinci Group Annual Report* (2009, p. 1).

B. Bouygues

In 2009, the ratio cash flow/revenue of Bouygues reached 10.9% (compared to Vinci at 15.3%). The Bouygues Group remained a profitable group, despite its pursuit of a very different strategy. At this time, in terms of the contribution by business area (see Table 2), the performance of Bouygues differed from that of the Vinci Group.

This was not always the case. At the end of 2003, public utilities represented 11.2% of Bouygues' turnover and 10.5% in the case of the Vinci Group. This was the consequence of the take-over of the SAUR Group in 1983. By diversifying, Francis Bouygues' strategy was to ensure regular and stable revenues (Barjot, 1992b). In 1982, he failed to repurchase the Drouot group, then France's second largest private insurer. One year later, he failed again in his attempt to obtain a majority shareholding of Lyonnaise des Eaux, then the Générale des Eaux. After these two failures, Bouygues focused his attention on the Société d'Aménagement Urbain et Rural (SAUR), which specialised in drinking water production, treatment and supply, providing utilities operations and management services to municipalities. Founded in 1933, SAUR underwent significant development during the 1960s. At this time, SAUR obtained important contracts in Nimes. It also created SODECI, a subsidiary in the Ivory Coast. At the beginning of 1980s, SAUR supplied over one million users and had sixteen regional directorates. At the same time, it pursued its international expansion, with a presence in the Congo, Guinea, Mali, Senegal and Zaire. From 1982, SAUR became established in Canada through its subsidiary, Aquatech. By 1983, SAUR constituted the third largest French group for water and urban sanitation improvement, following Compagnie Générale des Eaux (ranked at number one) and then Société Lyonnaise des Eaux et de l'Eclairage. SAUR generated 1.5 billion francs and supplied four million inhabitants (the Compagnie Générale des Eaux generated 30 billions francs and supplied 14 million inhabitants [see Lorrain, 1986]). Set up in rural municipalities, SAUR took advantage of the development of suburban areas. During the 1980s, it grew quicker than its two most important competitors.

One and a half years after entering the Bouygues group, SAUR had made a turnover of 2.7 billion francs. Thanks to new contracts of exploitation, management and services, SAUR was able to increase to 5000 units the number of municipalities and to 5 million the number of inhabitants to which it supplied. It developed engineering activities for the design and delivery of pumping and treatment plants of drinking water and sewage treatment plants. This was the beginning of a phase of rapid growth, to the detriment of its two major French competitors. SAUR obtained new contracts with French municipalities. It diversified its activity when it created

Table 2. Contribution by business area in 2009 (as a percentage of the total).

	Sales	Operating profit	Cash flow	Net profit
Bouygues Construction	29.3	17.5	17.0	15.6
Bouygues Immobilier (Real Estate)	9.5	10.6	5.4	7.1
Colas (Road Works)	36.7	28.3	31.9	24.3
TF1 (TV)	7.5	5.3	5.6	3.2
Bouygues Telecom	17.0	38.3	40.1	27.4
Alstom	0	0	0	22.4

Source: *Bouygues Annual Report* (2010, pp. 13–15).

Coved in France, the sanitation specialist of the group, and Ecovert, in the United Kingdom. Stereau penetrated the water treatment market and obtained construction contracts for a significant number of plants. Abroad, several subsidiaries were created: Sodeca (Guinea), Gestagna (Spain), SAUR UK (United Kingdom) and SNG (Poland). The Bouygues group also increased its interest in the energy sector. Thanks to SAUR, Bouygues inherited the ETDE (Entreprise de Travaux d'Equipement Electriques). Nevertheless the most spectacular diversification concerned the field of telecommunications.

During the years 1993–98, SAUR grew rapidly. From 10% of Bouygues' sales in 1993, its share increased to nearly 15% between 1995 and 1998. This was partly the result of a merger with Cise, one of its competitors, sold by the Saint Gobain Group in 1995. But, from 1999, SAUR's sales fell. Facing competition from Veolia Environnement and Lyonnaise des Eaux, Bouygues opted to avoid pursuing business in the less dynamic markets of water and sanitation. Above all, it was necessary to invest massively in the new telecommunications subsidiary, Bouygues Telecom. Consequently, in February 2005, Bouygues sold its shares of SAUR to PAI partners. Bouygues did not compete with Suez Environnement, Veolia Environnement or Vinci in the public utilities sector.

4. The second privileged sector: public utilities

In the French tradition, there has long been a strong link between public works and utilities and, therefore, between the two forms of concession (Bezançon, 2004b). The experience of Bouygues was no exception: GTM, Société de Construction des Batignolles and SGE were interested, at different times, in obtaining water concessions. In fact, the concession of public utilities constituted an important instrument of town planning (Curien & Dupuy, 1997; Desportes & Picon, 1997).

4.1. Public utilities as an instrument of town planning

Within the major branches of urban public utilities, there are three major trends as regards management: preponderance of private operators in water and urban waste, preponderance of public operators in the networks of energy, and a more complex situation in urban transportation (Lorrain, 2002).

A. Four dominant branches

Urban public utilities are divided into four major branches: networks of energy distribution; water distribution and collection; collection, treatment and valorisation of waste; and transport and parking services. During the recent period, these services grew at unequal rates as a consequence of citizens' needs and demands.

Whilst all branches grew due to increased sales, urban waste has constituted, in the recent period, the most dynamic of the four branches (Table 3). In contrast, value added increased slower than sales in urban waste and, particularly, in energy distribution. Most likely, in these two branches, there was the effect of decreasing returns to scale. Water seemed to be characterised by constant returns to scale and collective transportation by increasing returns to scale, perhaps because of the progress of interconnection between the different modes of transport. The consequence was a counterbalancing of the relative weight of different branches, particularly in terms of value added (Table 4).

In terms of sales only, the most important branch remained energy distribution whilst collective transportation took the biggest share of value added in 2006. Water distribution remained significant, whilst collective transportation suffered over the long term due to the amortisation burden. This was not the case for water distribution, which constituted a more interesting investment for private concessionaires of public utilities, such as in the case of the concession of underground parking (Barjot & Berneron-Couvenhes, 2005). With the exception of collective transportation and parking, in 2006, nearly all French citizens benefited from urban public utilities (Table 5).

Table 3. Average annual growth rates of urban public utility activity in France, 1997 to 2006 (percentage).

	Sales	Added value
Energy distribution	+4.2	+1.0
Water	+4.9	+4.8
Urban waste	+10.0	+8.5
Collective transportation*	+4.8	+7.3

Notes: *Parking excluded, subsidies included.
Source: Personal calculations from Ministère de l'Equipement, des Transports et du Logement (August 1999 and December 2008).

Table 4. Respective parts of sales and value added of the total of urban public utilities (percentages).

	Sales		Added value	
	1997	2006	1997	2006
Energy distribution	50.2	46.9	40.3	29.3
Water	21.2	20.8	24.5	24.8
Urban waste	8.4	13.2	8.2	11.6
Collective transportation*	20.2	19.1	27.0	34.3

Notes: *Parking excluded, subsidies included.
Source: Personal calculations from Ministère de l'Equipement, des Transports et du Logement (August 1999 and December 2008).

Table 5. Population benefiting from urban public utilities in 1997 and 2006 (millions of inhabitants).

	1997	2006	Average annual growth rate (percentage) 1997–2006
Energy distribution	44.7	52.9	+1.9
Water	44.7	52.9	+1.9
Urban waste	44.7	52.9	+1.9
Collective transport	32.2	33.7	+0.5

Notes: *Parking excluded, subsidies included.
Source: Personal calculations from Ministère de l'Equipement, des Transports et du Logement (August 1999 and December 2008).

Between 1997 and 2002, growth was fastest in energy distribution, water and urban waste, after which, it sharply decreased. In contrast, collective transportation grew from 2002 to 2006 after a phase of net decline. Moreover, collective transportation remained the most important employer in 2006 (118,700 employers), ahead of energy distribution (106,200 employers), urban waste (60,600) and water distribution (45,700). Between 1997 and 2006, water distribution, urban waste and, particularly, energy distribution, created more new jobs.

The weakness of depreciation charges (in water and urban waste) and the possible labour productivity gains (in urban waste) or, at least, their stability (in water distribution) favoured constant interest on the part of private firms in these two branches (Table 6).

B. Water and urban waste: two attractive branches for private capital

Water distribution constituted the most successful case of the French model of concessionaire capitalism, with the archetypal examples of Compagnie Générale des Eaux and Lyonnaise des Eaux (Bonin, 2005). Indeed, in 2006, private operators were dominant (Table 7).

Concerning water treatment and distribution, private dominance was clear. But, from 1997, the relative weight of private operators also grew in cleansing and filtering. Water constituted a good example of narrow oligopoly.

By 2006, around 70% of the market was controlled by the three most important groups (Table 8): almost 40% by Veolia Eau, subsidiary of Veolia Environnement Group; a little over 20% by Lyonnaise des Eaux France, subsidiary of Suez-Ondéo, a big branch of Suez Environnement; and around 10% by SAUR France, subsidiary of SAUR. Veolia Environnement appeared on the defensive facing two independent

Table 6. The workforce growth and labour productivity increases in urban public utilities from 1997 to 2006 (average annual growth rates in percentages).

	Growth of workforce	Gains of annual labour productivity[1]
Energy distribution	+6.9	−5.9
Water	+4.9	−0.1
Urban waste	+6.3	+2.2
Collective transportation	+2.1	+5.2

Notes: [1]Added value by employee.
Source: Personal calculations from Ministère de l'Equipement, des Transports et du Logement (August 1999 and December 2008).

Table 7. The role of the private sector in water distribution from 1997 to 2006 (percentage of total and by profession).

Treatment and distribution	80	81
Private operators	20	19
Public operators		
Cleansing and filtering	55	59
Private operators	45	41
Public operators		

Source: Personal calculations from Ministère de l'Equipement, des Transports et du Logement (August 1999 and December 2008).

operators, but SAUR France had the same attitude. The position of public operators remained stable and significant, comprising almost one third of the market.

Urban waste constituted a second attractive branch to private investors. Nevertheless, the degree of private intervention varied according to the different categories (Table 9).

Concerning collection operations, the ratio private/public reached 60:40, whilst the role of private operations continued to grow. On the contrary, private operators completely dominated in treatment and valorisation, even in those sectors where the activity of public operators was increasing. Private operators were generally subsidiaries of three big water distribution groups (Table 10).

Table 8. Distribution of the French water market between the most important operators (percentage of the served urban population).

	2000	2006
1/ Private operators	71	71
- Veolia Environnement (Ex. Générale des Eaux)	42	37
- Suez Environnement (Ex. Lyonnaise des Eaux)	19	21
- Saur (Bouygues, then PAI partners)[a]	10	11
- Independent	0	2
2/ Public operators (municipal agencies, intercity syndicates)	29	29

[a]Today controlled by the Caisse des Dépôts, Séché environnement and AXA Investissement.
Source: Personal calculations from Ministère de l'Equipement, des Transports et du Logement (August 1999 and December 2008).

Table 9. The role of the private sector in urban waste branches from 1997 to 2006 (percentage of total of treated quantities and 2006 by category).

	1997	2006
1/ Collection:	60	64
- private operators	40	36
- public operators		
2/ Treatment and valorisation:	92	90
- private operators	8	10
- public operators		

Source: Personal calculations from Ministère de l'Equipement, des Transports et du Logement (August 1999 and December 2008).

Table 10. Distribution of the French urban waste market between the most important operators (percentage of the managed quantities).

	2000	2006
1/ Private operators	42	34
-Veolia Environnement	24	20
-Suez Environnement	10	10
-Saur	0	3
-Independent		
2/ Public operators (municipal agencies, intercity syndicates)	28	33

Source: Personal calculations from Ministère de l'Equipement, des Transports et du Logement (August 2002 and December 2008).

Despite the progress of independent operators, two major facts were the domination of a narrow oligopoly of large private operators and the strong opposition of public operators. This was the result of a disengagement of Veolia Propreté, subsidiary of Veolia Environnement, and SITA, subsidiary of Suez Environnement. In contrast, Coved, subsidiary of SAUR, maintained its position.

4.2. Two historical models: Générale des Eaux versus Lyonnaise des Eaux

Competition between Véolia Environnement (see Véolia Environnement, 2003) and Suez Environnement resulted in two opposite historical models of development, according to specific process of path dependency.

A. Générale des Eaux: a leader in water public utilities

The Compagnie Générale des Eaux was founded during the Second Empire on 24 December 1853 (Jacquot, 2002). Its founding partly relied on contributions from French and British capital by Count Henri Siméon (1803–74), former prefect and General Director at the French Ministry of Finance, sustained by the French Haute Banque (High Bank).[1] The company had two purposes: to irrigate the French countryside and supply water to cities. Siméon and Générale des Eaux directors wished to obtain the concession of water distribution of Paris, with a statute of public company. For this reason, they listed the company on the Bourse de Paris in 1854 as a national and international venture.

It was impossible, however, to obtain a concession in Paris until 1860 when a statute on managing the interests of the public water service was created. In compensation, Compagnie Générale des Eaux acquired immediately, in 1853 and 1854, many important concessions in the suburbs of Paris, Nantes and, particularly, Lyon. Because of the sound results of the Parisian experience, in 1867, the state authorised the company to develop its activities in Parisian suburbs. During the Second Empire, the firm set up in the South East of France (in Nice, and suburbs of Lyon). In 1879, the successor to Count Siméon, Sir Edward Blount, a British Catholic banker, created a second public company, the Compagnie Générale des Eaux pour l'Etranger (CGEE). The sister company of CGE grew rapidly, obtaining many foreign concessions (in Venice in 1880, Constantinople in 1883, and Porto in 1883). In 1884, CGE diversified its skills to the treatment of waste water. Consequently, in 1889, the company opened a central laboratory in Paris. The most important result of the laboratory was the replacement of chlorine by ozone gas in water filtration and sterilisation.

Despite the difficulties during the First World War, CGE was re-launched during the interwar period, thanks to local cooperation with Lyonnaise des Eaux, acquisition of new concessions in Champagne, Nord, and Normandy, and the establishment of two subsidiaries specialised in public works. So, the company was well positioned to resist the difficult economic context of the 1930s and during the Second World War. The management remained prudent and even conservative. Two main directors were responsible: Georges Huvelin (1907–76), chief executive officer in 1952, then chairman in 1972, and his successor, Guy Dejouany (Barjot, 2010).

Under their respective direction, Générale des Eaux became a powerful group. From 1956, CGE took a part in the capital of a big French contractor, Campenon

Bernard (Barjot, 2008). In 1966, Générale des Eaux became the second most important shareholder after the Banque de l'Union Parisienne, then Paribas. This was the beginning of a battle for leadership in the construction sector between this company and Bouygues (Barjot, 1992a). Générale des Eaux was also interested in energy distribution (taking over Compagnie Générale des Chauffe in 1967, and then establishing Dalkia in 1998), urban waste (creating Onyx in 1989 via a merger between a group of specialised firms) and the conquest of foreign markets (taking over US Filter in the United States and FCC in Spain, respectively, in 1990 and 1998).

From 1984, Générale des Eaux diversified into television (Canal Plus in 1984), then telecommunications (SFR in 1998). In 1996, Jean-Marie Messier succeeded Guy Dejouany as chairman and chief executive officer. He wanted to become a world leader in media and communications. Consequently, Compagnie Générale des Eaux became Vivendi (1998). Despite the acquisition of Havas (1998) and an alliance with Mannesmann against Vodafone (1999), the new strategy resulted in a succession of failures. The company had to compete with Lyonnaise des Eaux and Suez (1997) for the takeover of the Canadian Seagram Group in order to control Universal (2000). Then, Vivendi merged with Seagram and Canal Plus (2000). These activities constituted an excessive and dangerous diversification. In 2002, the new Vivendi Universal Group announced enormous losses (13.6 billion euros).

A new history began with the independence of Vinci (*vide supra*) and Vivendi Environnement. From 1999, Vivendi merged all its activities related to environmental services: Vivendi Water, Onyx (sanitation), Dalkia (energy services) and Connex (collective transportation). Named Vivendi Environnement, the new company, was listed on the Paris Stock Exchange (July 2000), then on the NYSE (October 2001). In the same year, 2000, Vivendi Environnement concluded an alliance with EDF. Rapidly, the firm gained its financial independence, becoming Veolia Environnement from 2003.

B. Lyonnaise des Eaux: a world group of services of energy and sustainable development (Beltran, Williot, Moreau, Brion, & Bonin, 2008)

Founded on an initiative of Crédit Lyonnais in February 1880, the Société Lyonnaise des Eaux et de l'Eclairage (SLEE) rapidly became the first rival of Générale des Eaux (Goubert, 1986). SLEE obtained a good number of important concessions in France, including those in Toulon (1884), Dunkerque (1892) and Bordeaux (1906), but also in foreign countries, including Spain (Société des Eaux de Barcelone in 892) and Morocco (Société Marocaine de Gaz et d'Electricité in 1914). At the same time, Lyonnaise des Eaux played a major role in the development of the gas industry and in the process of electrification across French territory. Under the direction of Albert Petsche (1860–1933), then Ernest Mercier (1878–1955), SLEE developed an active strategy, alternating between internal and external growth: in the gas industry, mergers with Compagnie Française du Centre et du Midi (1914), Union des Gaz (1929) and Compagnie Gazière Européenne (1935); and in the electricity sector, the constitution of Union d'Electricité (1919), Union Financière pour l'Industrie et l'Electricité (UNIE in 1930) and Union Electrique de l'Outre-mer (UNELCO in 1935).

The nationalisation of the gas and electricity industries in France compelled the company to adopt a new strategy. Henceforth named Société Lyonnaise des Eaux, it diversified into other sectors, such as electric works (Forclum in 1947, Trindel in 1972), pipe-lines equipment (COFOR in 1957), road works (Cochery in 1947) and construction (Maisons Phénix in 1961). Always well-established in the water distribution market, thanks to the central laboratory of Le Pecq (1933), in the Parisian Region, Lyonnaise des Eaux took control, in 1972, of Degrémont, a world leader of water treatment engineering, and took a part in SITA, a powerful company in charge of urban waste treatment. Between 1967 and 1974, Crédit Lyonnais sold the control of Lyonnaise des Eaux to the Compagnie de Suez, becoming the major shareholder. In 1980, Jérôme Monod became Chairman and CEO of the Lyonnaise des Eaux (Barjot, 2010). Former chief executive officer of the Délégation à l'Aménagement du Territoire (a powerful public agency in charge of the French planning territory) and general secretary of the Gaullist party (Rassemblement Pour la République [RPR]), Monod adopted an efficient strategy, which consisted of reducing interests in cyclical activities (construction and real estate with GTM and Dumez) and developing service activities: communications (creating M6, a TV channel) and environment (taking over Elyo in 1994).

In 1997, Monod convinced Gérard Mestrallet, chairman and CEO of the Compagnie Financière de Suez, to merge their two groups. The result was the constitution of the second most important French group at that time. Consequently, Suez operated important divestments in construction (Vinci), in television (TPS in 2002, Paris Première in 2003, and M6 in 2004) and telecommunications (Noos). At the same time, the group reinforced its strategic positions in the environment sector, including a takeover bid of SITA (1998), acquisition of United Water (USA in 1999), the establishment of Suez Environnement after a merger between Degrémont, Ondéo and SITA, and a takeover bid of Agbar (in Spain in 2008). The greatest successes were realised in the energy sector with the takeover of Electrabel (Suez-Tractebel in 2003) and, above all, the merger of Suez-GDF, the biggest competitor of EDF in the European energy sector.

5. Conclusion: the rise of urban corporations and of a market for infrastructures

The rise of powerful urban corporations specialised in infrastructure and public utilities constitutes a major characteristic of our time (Barjot, 2003b). This is a result of the growing place of cities in contemporary capitalism and the change of the urban growth model around the development of complex technical networks and new services. This article has argued that the French legal system stimulated 'network capitalism' by establishing a financial and managerial system which allowed for the participation of private capital in the development of major infrastructure projects. This gave French companies a particular advantage when they internationalised, most particularly, from the 1990s.

It is possible to identify seven major types of actors in urban infrastructure: water operators (generally public), waste treatment operators (also, mostly public), electricity (Chick & Nelles, 2007), engineering, builders, capital goods producers and, above all, big multi-utilities groups such as RWE and EON in Germany, Enel in Italy, Mitsui and Mitsubishi in Japan and Cheung Kong Holdings in China (Hong Kong). France produced four important groups in this latter category: Bouygues, Vinci, Veolia Environnement and GDF-Suez. These firms belonged to a sort of 'network capitalism' and shared the fact that they were subject to heavy sunk costs, concluded contracts with public utilities and executed public service missions. As in the case of other world competitors, these French firms concluded long-term contracts, a necessary condition for sufficient profits. Despite evident failures, for instance, in communication networks, the international competitiveness of these French groups was the result of a long-term tradition of interest on the part of engineers in networks systems (Ecole nationale des Ponts et Chaussées, Ecole Centrale de Paris), questions relating to energy (Ecoles nationales des Mines), but also due to the intervention of French banks such as Suez or Paribas in major international ventures. Building Suez was a quasi-perfect example of this French experience, because it was the result of three successive major operations: the takeover of the Société Générale de Belgique (1988), the merger of Suez-Lyonnaise des Eaux (1997), and the merger of GDF-Suez (Bonin, 2010).

Note

1. Siméon married a daughter of the French banker François-Alexandre Seillière. See Stoskopf (2002).

References

Barjot, D. (1988). From Tournon to Tancarville. The contribution of French civil engineering to suspension bridge construction, 1824–1959. *History and Technology*, *6*, 177–201.

Barjot, D. (1989). *La grande entreprise française de travaux publics (1883–1974). Contraintes et stratégies* (Unpublished doctoral thesis). Paris-Sorbonne University (Paris IV), France.

Barjot, D. (1990). Un grand ouvrage du XXe siècle: Le pont de Tancarville. In D. Woronoff (Ed.), *L'Homme et l'industrie du néolithique à nos jours* (pp. 311–323). Alençon: Société historique et archéologique de l'Orne.

Barjot, D. (1992a). *Fougerolle. Deux siècles de savoir-faire*. Paris: Editions du Lys.

Barjot, D. (1992b). L'ascension d'un entrepreneur: Francis Bouygues (1952–1989). *XXe siècle*, *35*, 42–59.

Barjot, D. (1993). *Les travaux publics de France: Un siècle d'entrepreneur et d'entreprises. 1883–1992*. Paris: Presses de l'Ecole Nationale des Ponts et Chaussées.

Barjot, D. (1994). Les entrepreneurs français de travaux publics et l'équipement du Portugal: une contribution multiforme (milieu du XIXe siècle-milieu des années 1970. *LER-Historia*, *26*, 93–116.

Barjot, D. (1995). Les grandes entreprises européennes de travaux publics face au marché international (1880–1914). *HES*, no. 2, 361–383.

Barjot, D. (2003a). *La trace des bâtisseurs: Histoire du Groupe Vinci*. Paris: Vinci.

Barjot, D. (Ed.). (2003b). *Entreprises et Histoire* [Special issue – Globalization-la mondialisation], *32*(1).
Barjot, D. (2003c). L'entreprise Hersent: Ascension, prospérité et chute d'une famille d'entrepreneurs (1860–1982). In J.-C. Daumas (Ed.), *Le capitalisme familial: Logiques et trajectoires* (pp. 133–160). Besançon: Presses Universitaires franc-comtoises.
Barjot, D. (2006a). *La grande entreprise française de travaux publics (1883–1974)*. Paris: Economica.
Barjot, D. (2006b). Les nationalisations de la Libération. In C. Stoffaes (Ed.), *Psychanalyse de l'antilibéralisme. Les Français ont-ils raison d'avoir peur?* (pp. 46–156). Paris: Editions Saint-Simon (IDHI).
Barjot, D. (Ed.). (2007). *Revue économique* [Special issue – Où va l'histoire des entreprises?], *58*(1)
Barjot, D. (2008). Introduction générale. In D. Barjot and J. Dureuil (Eds.), *150 ans de génie civil: Une histoire de centraliens* (pp. 15–46). Paris: Presses de l'Université Paris-Sorbonne.
Barjot, D. (2009a). Nationalisation et dénationalisation: Une mise en perspective historique. Quand la crise transcende: Les axes idéologiques. *Revue Economique et Sociale*, *67*, 13–28.
Barjot, D. (2009b). Skanska (1887–2007): The rise of a Swedish multinational company. *Nordic Historical Review*, *8*, 225–256.
Barjot, D. (2010). Famille Bouygues. In J.C. Daumas (Ed.), *Dictionnaire historique des patrons français* (pp. 122–126). Paris: Flammarion.
Barjot, D., & Berneron-Couvenhes, M.F. (Eds.). (2005). *Entreprises et Histoire* [Special issue: Concession et optimisation des investissements publics], *38*(1).
Barjot, D., & Dureuil, J. (Eds.). (2008). *150 ans de génie civil: Une histoire de centraliens*. Paris: Presses de l'Université Paris-Sorbonne.
Barjot, D., Morsel, H., & Badel, L. (Eds.). (1996). *La nationalisation de l'électricité en France. Nécessité technique ou logique politique?* Paris: Presses Universitaires de France.
Barjot, D., Petitet, S., & Varaschin, D. (Eds.). (2002). *Entreprises et Histoire* [Special issue – La concession, outil de développement], *31*(3).
Beltran, A., Chick M., & Lanthier P. (Eds.). (2003). *Annales historiques de l'électricité* [Special issue – Nationalisations et dénationalisations de l'électricité], *1*(June).
Beltran, A., Williot, J.P., Moreau, J.-L., Brion, R., & Bonin, H. (2008). *Suez. L'Histoire*. Paris: Editions PC.
Berneron-Couvenhes, M.F. (2007). *Les messageries maritimes. L'essor d'une grande compagnie de navigation française, 1851–1894*. Paris: Presses de l'Université Paris-Sorbonne.
Berthereau, D. (2005a). *L'expertise de la Commission de vérification des comptes des entreprises publiques (1948–1976)* (Unpublished doctoral thesis). Paris-Sorbonne (Paris-IV) University, Paris, France.
Berthereau, D. (2005b). Le contrôle d'une entreprise concessionnaire par la commission de vérification des comptes des entreprises publiques: Le cas du tunnel du Mont-Blanc (1965–1971). *Entreprises et Histoire*, *38*, 84–95.
Bettinger, C. (1997). *La gestion déléguée des services publics dans le monde; concession ou BOT*. Paris: Berger-Levrault.
Bettinger, C., & Le Chatelier, G. (1995). *Les nouveaux enjeux de la concession après les réformes de 1993–1995*. Paris: EFE.
Bezançon, X. (1999). *Essai sur les contrats de travaux et de services publics; contribution à l'histoire administrative de la délégation de mission publique*. Paris: Librairie Générale de Droit et de Jurisprudence.
Bezançon, X. (2004a). *2000 ans d'histoire du partenariat public-privé pour la réalisation des équipements et services collectifs*. Paris: Presses de l'Ecole Nationale des Ponts et Chaussées.
Bezançon, X. (2004b). *Les services publics en France* (Vol. 1 – *Les services publics du Moyen-âge à la Révolution*; Vol. 2 – *Les services publics en France de la Révolution à la première guerre mondiale*). Paris: Presses de l'Ecole Nationale des Ponts et Chaussées.
Bezançon, X. (2005). Aux origines du droit concessionnaire en France. In D. Barjot & M.F. Berneron-Couvenhes, Concession et optimisation des investissements publics [Special issue], *Entreprises et Histoire*, *38*, 24–54.
Bonin, H. (2005). Le modèle français du capitalisme de l'eau dans la compétition européenne et mondiale depuis les années 1990. In C. Baron (Ed.), Société civile et marchandisation de l'eau. Expériences internationales. *Sciences de la Société*, *64*, 55–73.

Bonin, H. (2010). *History of the Suez Canal Company, 1858–1960. Between controversy and utility*. Paris: Droz.
Bouygues Annual Report. (2010). Paris: Bouygues.
Chevallier, J. (1987). *Le service public*. Paris: Presses Universitaires de France.
Chick, M., & Lanthier, P. (Eds.). (2004). *Entreprises et Histoire* [Special issue – Nationalisations et denationalisations], *37*(3).
Chick, M., & Nelles, H.V. (2007). Nationalisation and privatisation. Ownership, markets and the scope for introducing competition into electricity supply industry. In D. Barjot (Ed.), Où va l'histoire des entreprises? *Revue économique*, *58*(1), 277–293.
Conchon, A. (2005). Financer la construction d'infrastructures de transport: La concession aux XVIIe et XVIIIe siècles. *Entreprises et Histoire*, *38*, 55–70.
Curien, N., & Dupuy, G. (Eds). (1997). *Réseaux de communication: marchés et territoires*. Paris: Presses de l'ENPC.
Daumas, J.C. (Ed.). (2010). *Dictionnaire historique des patrons français*. Paris: Flammarion.
Desportes, M., & Picon, A. (Eds.). (1997). *L'espace au territoire: L'aménagement en France*. Paris: Presses de l'ENPC.
Foreman-Peck, J., & Milward, R. (1990). *Public and private ownership of British industry 1820–1990*. Oxford: Oxford University Press.
Goubert, J.P. (1986). La rente de l'eau. La stratégie industrielle de la Société Lyonnaise des Eaux et de l'Eclairage. 1880 – 1925. *Annales de la Recherche Urbaine*, *30*, 131–152.
Jacquot, A. (2002). La Compagnie Générale des Eaux 1852–1952: Un siècle, des débuts à la renaissance. *Entreprises et Histoire*, *30*, 32–44.
Lorrain, D. (1986). L'industrie des réseaux urbains en France: des origines nationales à une dynamique mondiale. *Revue d'économie régionale et urbaine*, 1, 95–109.
Lorrain, D. (Ed.). (2002). *Entreprises et Histoire* [Special issue – Les grands groupes et la ville], *30*(2).
Lorrain, D., & Stoker, G. (Eds.). (1995). *La privatisation des services urbains en Europe*. Paris: Editions La Découverte.
Marnot, B. (2005). Les ingénieurs parlementaires de la Troisième République et les figures respectives de la concession. *Entreprises et Histoire*, *38*, 71–83.
Martinand, C. (Ed.). (1993). *L'expérience française du financement privé des équipements publics*. Paris: Economica.
Massé, P., & Bernard, P. (1969). *Les dividendes du progrès*. Paris: Le Seuil.
Mescheriakoff, A.S. (1991). *Droit des services publics*. Paris: Presses Universitaires de France.
Millward, R. (2005). *Private and public enterprise in Europe: Energy, telecommunications and transport c. 1830–1990*. Cambridge: Cambridge University Press.
Ministère de l'Equipement, des Transports et du Logement. (1999, August). *Les opérateurs privés et publics de services urbains en France en 1997*. DAEI.
Ministère de l'Equipement, des Transports et du Logement. (2002, August). *Les opérateurs privés et publics de services urbains en France en 2000*. DAEI.
Ministère de l'Equipement, des Transports et du Logement. (2008, December). *Les opérateurs privés et publics de services urbains en France en 2006*. DAEI.
Montel, N. (1998). *Le chantier du canal de Suez, 1859–1869. Une histoire des pratiques techniques*. Paris: Presses de l'Ecole Nationale des Ponts et Chaussées.
New Engineering Records. (2009). The 2009 top 225 international contractors and top 225 global contractors. Retrieved from http://www.cmc.coop/img/articoli/536/ENR-09-TIC.pdf
Park, R.R. (2000). La Société de Construction des Batignolles: Des origines à la première guerre mondiale (1846–1914): Premiers résultats. *Histoire, Économie et Société*, *3*, 361–386.
Park, R.R. (2005). *La Société de Construction des Batignolles. Des origines à la Première Guerre Mondiale (1846–1914)*. Paris: Presses de l'Université Paris Sorbonne.
Perrot, J.Y., & Chatelus, G. (2000). *Financement des infrastructures et des services collectifs. Le recours au partenariat public-privé*. Paris: Presses de l'Ecole Nationale des Ponts et Chaussées.
Petitet, S., & Varaschin, D. (1999). *Intérêts publics et initiatives privées, initiatives publiques et intérêts privés. Travaux et services publics en perspectives*. Vaulx-en-Velin: Ecole Nationale des Travaux Publics de l'Etat.

Piquet, C. (2008). *La Compagnie du Canal de Suez. Une concession française en Egypte (1888–1956)*. Paris: Presses de l'Université Paris-Sorbonne.
Piquet, C. (2009). *Histoire du canal de Suez*. Paris: Perrin.
Reverdy, G. (2004). *Les travaux publics en France (1817–1847). Trente années glorieuses*. Paris: Presses des Ponts et Chaussées.
Saul, S. (1997). *La France et l'Egypte de 1882 à 1914. Intérêts économiques et implications politiques*. Paris: Comité pour l'Histoire Economique et Financière de la France.
Schneier, G., & de Gouvello, B. (Eds.). (2003). *Eaux et réseaux. Les défis de la mondialisation*. Paris: CREDAL-IHEAL Editions.
Stoskopf, N. (2002). *Les patrons du Second Empire. Vol. 7 – Banquiers et financiers parisiens*. Paris: Picard-Cénomane.
Toninelli, P.A. (Ed.). (2000). *The rise and fall of state-owned enterprises on the Western world*. Cambridge: Cambridge University Press.
Véolia Environnement. (2003). *Véolia Environnement 1853–2003. 150 ans au service de l'environnement*. Paris: Editions Cercle d'Art.
Vinci Group Annual Report. (2009). Rueil-Malmaison: Vinci Group.

Index

Page numbers in *Italics* represent Tables and page numbers in **Bold** represent Figures